THE NORTH AMERICAN
MARIA THUN
BIODYNAMIC
ALMANAC
2021

CREATED BY

MARIA AND MATTHIAS THUN

Floris
Books

Compiled by Matthias Thun
Translated by Bernard Jarman
Additional astronomical material
by Wolfgang Held and Christian Maclean

Published in German under the title *Aussaattage*
English edition published by Floris Books

MIX
Paper from
responsible sources
FSC® C117931

British Library CIP Data available

ISBN 978-178250-653-9
ISSN: 2052-577X

Printed by MBM Print SCS Ltd, Glasgow

Floris Books supports sustainable forest management by
printing this book on materials made from wood that comes
from responsible sources and reclaimed material

Walter Thun, Flowering Willow in Willingshausen, *(watercolor, 12 x 16 in)*

The dynamic quality of this simple and eye-catching painting of a willow is enhanced by the diagonal lines of the branches meeting those of the riverbank.

Introduction

The planetary effects in the monthly notes for 2019/2020, were as expected. Not only was it difficult for plants to cope with the drought in many parts of the world, the lack of water reduced the nectar flow in the flowers, and there was, for example, virtually no honeydew. This limited the swarming drive among bee colonies too. The forests also struggled with the drought; it hindered the formation of tree resin and made the trees more susceptible to numerous pests. For bees the year was underwhelming, while for many a forester and woodland farmer the year was a shock. For 2020/2021 we therefore hope for more rain.

Easter 2019

Matthias Thun

During 2019 we received many questions concerning the date of Easter. We had in our calendar simply followed the accepted date of Easter as April 21. Easter is the first Sunday after the first Full Moon after the spring equinox. In practice it is calculated by a series of tables devised by the mathematician Carl Friedrich Gauss in the early nineteenth century, and assumes the spring equinox to be on March 21.

However, in 2019 the spring equinox was actually on March 20 (at 9:58 pm GMT), and if we strictly follow the astronomical rule rather than the tabular calculation, the first Full Moon was on Thursday March 21 at 1:42 am (GMT), and so Easter would be on Sunday March 24.

To make matters more complicated, we could of course ask, why should the time (and date) of Greenwich be used? Should it not be in Jerusalem? Using the local time of Jerusalem, the equinox would indeed be on March 21 (just after midnight). Should then the Full Moon occurring on the same date be used, or should it be the Full Moon on a date *after* the equinox?

In short, the date of Easter in 2019 was ambiguous. Which date would actually affect the world of plants and animals? We sought some guidance from the climatic phenomena but unfortunately to no avail. In previous years it was possible to recognize the transition from Good Friday to Easter Sunday through, for example, the behavior of birds. In 2019 however, neither of the two possible dates gave a definitive indication.

We therefore decided to carry out comparative trials around both dates. The plants we used were spring oats, spring rye, tomatoes, cucumbers, peppers and radishes.

As the chosen dates approached a few minor problems presented themselves. That spring, which had begun quite mild, was not conducive to steady and even plant growth. Right up until March 20 we had repeated night frosts. During the first trial period from March 21 to 24 it was sunny but the night frosts and a light sprinkling of snow prevented any outdoor sowing. In order to germinate vegetable plants require an ongoing night-time temperature of at least 36–37°F (3–4°C). If frost interrupts the process of germination it will stop and in a subsequent warm period germination will not occur evenly because the seeds cannot repeat their initial push to germinate.

All our sowings therefore had to take place in the greenhouse. This was no problem for the vegetable seeds and plants. Only for the cereals, which are normally sown directly where they are to mature, did we have to find another solution. They were sown in larger box containers that could be planted out four weeks later during the second trial period.

In the second period the cereals were sown directly outside and the vegetables were sown in the greenhouse and transplanted outside later. Sowing the various seeds in spring under the same conditions at an interval of four weeks with varying weather was clearly no easy matter.

Since we had to reckon with a spring that did not progress harmoniously and the Easter sowing dates were fixed, we had at least to ensure there was sufficient frost-free soil available. We therefore calculated the amount of soil needed and prepared a mix.

Soil was needed for the greenhouse bed and for the plant pots and boxes. To make a good mixture one third was well rotted, two-year-old cow manure compost, one third meadow soil and one third organic seed compost that was bought in. These were thoroughly mixed and in the process repeatedly treated with barrel preparation (cow pat pit, CPP) so that the soil from outside could benefit from the compost preparations.

The soil mix was left for about four weeks so that its three components could form a single entity. Because the barrel preparation leads to a good composting and transformation process, it is essential that horn manure is applied when the seeds are sown in order to halt the rotting and fungal processes. Without the horn manure there is a danger that the composting processes in the soil will transfer to the plants and cause fungal problems.

Horn manure preparation should be sprayed three times. The first application should be made in the appropriate constellation (for instance, Fruit times for tomato). Then the cosmic effect is strengthened and is more visible in the plants. In biodynamic agriculture horn silica is normally sprayed at the appropriate stage of growth to improve quality. In these particular trials we have refrained from using it because the silica application could mask the effects of the Easter constellations.

Previous trials had shown that the effects would only be seen in the next plant generation. This means collecting seeds from the cereals and tomatoes, for example, and growing them the following year. The observations and experiences of previous years lead us to hope that similar observations could be made again.

Friedrich with the single grains. Sowing vegetable seeds during the first Easter dates.
Left: Maundy Thursday, right: Good Friday
Later it was found that the first seeds sown in a taller plant pot thrived better before being pricked out.

Titia sows oats in tubs where they produced fine deep roots.

Friedrich sowing oats in the greenhouse bed.

Friedrich sowing tomatoes in the greenhouse.

Horn manure is applied to the seed beds to ensure that the soil composting process set in motion by the barrel preparation is ended.

We have found time and again that the days immediately preceeding Easter tend to be somewhat oppressive and that on Easter Sunday the tension is released. On Easter Sunday at dawn the birds suddenly come out in full song. I was outside at dawn during both of the trial 'Easter' periods but this year the birds remained silent. On the second Easter date however two wagtails showed by their delightful pairing behavior, that something had changed in their cycle of life.

In conversations with various friends about the two Easter dates, some of whom had also carried out sowing trials and wanted to experience the mood, I found that similar experiences had been made. Significant differences were, however, also presented which could be traced back to the Moon constellations of those days or else to feelings that arose.

Whether we will succeed with the help of the plants and their subsequent progeny in finding the answer to the engima of the date of Easter in 2019 will only be clear after two growing seasons. We have come to realize over many years that we cannot rely solely on weather phenomena, but that only through the next plant generation can there be any certainty. The resulting harvest can therefore only be reported on in the next calendar.

Stirring the preparations. We stir small quantities in cylindrical earthenware pots.

Horn manure is also applied to the seeds sown into the greenhouse soil.

More on Planets and Trees

In reflecting on the 2020 edition of the *Maria Thun Biodynamic Almanac,* we would like to expand upon what was presented regarding the connection between trees and planets and the reference made by Rudolf Steiner to the way planetary forces are expressed in fruiting trees.

While woody plants are of course also dependent on the Sun they have a much stronger connection to the planets in our solar system. In the Agriculture Course Rudolf Steiner refers to the connection of conifers to Saturn and the oak tree to Mars. Later on, in the second lecture, he said: 'Just like the color of flowers, the fine taste of apricots or plums is a cosmic quality that has made its way up into the fruit. In every apple you are actually eating Jupiter, in every plum, Saturn.' (p. 40).

In other places he describes the living relationship between the tree and the forces of the stars. In the case of woody plants we often find in the fruit the planetary influences expressed from the previous year. This has led us and some of our readers to ask whether a fruit-like quality can be found in shrubs that don't bear edible fruits.

When working with different types of wood we soon discover that they emit wonderful smells. These scents often help overcome feelings of indisposition associated with many an ailment. We have also noted elsewhere how a strong flow of nectar can be detected in the evening air beside beehives as the nectar is being transformed to honey. This scent has a crystal-like quality, strengthening our innermost self.

When describing evolution Rudolf Steiner describes how during the Atlantean epoch the foundation of the human self and self-awareness was being developed in parallel with the formation of the six-sided crystal – an amazing complementary creative process.

Maples

The living fluid of the maple, the blood of the tree, is a resource that can be tapped at certain times of year and thickened into maple syrup.

It is not honey that is formed from the nectar of flowers, but the sieved liquid sap containing about 3% sugar which then has to undergo an elaborate process to produce a thickened syrup. The maple honey originating from the nectaries of the flowers (all flowers have them) is of a higher value since it is produced in the nectaries of flowers during certain planetary aspects.

Teas can also be made from the flowers, enabling Jupiter forces to be taken up. Passionate herb tea drinkers will, however, miss a strong flavor. The taste of maple tea is very subtle and unassuming. It can however be strengthened with a 'taste enhancer' in the form of flower honey. Only a very small amount of honey

A red and a yellow maple tree. In spring the glistening pale color of the yellow maples is particularly striking. They produce so much nectar that other flowers which bloom at the same time such as rape are often neglected by the bees.

Red maple flowers

Yellow maple flowers

Maple flower buds

Green, unripe maple seeds

should be added – not more than ¼ teaspoonful per cup. Other forms of sugar or maple syrup should be avoided since it is only maple honey which can bring out the flavor. Beet sugar should likewise be avoided since its earthy, root qualities overpower the fine nuances of flavor. This also applies to the blossom teas described below.

Elder flowers

Elderberries

Elderflower tea. The flower teas can be made from fresh or dried flowers. A good tasting syrup can be produced from elderberries. Once they are ripe the berries need to be harvested quickly before the birds eat them.

The First Nine Years of Research

Maria Thun

Introduction by Matthias Thun

By way of an introduction and to make it easier to understand the very difficult texts written by Maria Thun about the first nine years of research, some background will be given regarding the historical context out of which the decision was made to investigate, understand and verify influences from the cosmos.

In the immediate aftermath of the Second World War an attempt was made in many fields of life to rethink the thoughts and ideas inherited from the past. Rudolf Steiner's Agricultural Course had taken place in 1924 and the questions which had arisen in relation to agriculture before the war could now be taken up and put into practice. Rudolf Steiner placed much emphasis on the importance of enlivening the soil with the help of special preparations and on the influence of the cosmos on living processes.

The old farming rules inherited from the past still lived strongly in the minds of farmers and gardeners and it was believed that new astronomical questions could find answers and build on this old knowledge. It very quickly became apparent, however, that the old farming rules were no longer effective and often appeared to be false. It is worth pointing out that those rules were not wrong; it was just that the constellations with which they had been connected, were not known.

Maria Thun knew from her parents that in their youth the many tasks in the garden and on the fields determined by weather phenomena were carried out at particular times. Having learnt from Rudolf Steiner about the close connection between plant life and the cosmos she began the experiments that are described in the following text. These initial trials laid the foundation for the later successive planting trials that are being continued on our farm to this day.

Sowing with Moon phases and nodes

In the Agriculture Course Rudolf Steiner pointed out very specific areas where plant growth is influenced by the planets. He pointed particularly to the forces of the Full Moon and how they work beneficially on the growth of young seedlings. These indications were taken up by many people engaged with biodynamic agriculture and put into practice. In this way Franz Rulni together with Heinrich Schmidt produced a sowing calendar that focused particularly on the phases, nodes, perigee and apogee and on the ascending and descending cycles of the Moon.

In applying the recommendations of Franz Rulni in 1952 I made the following observations. Within one Full Moon period there were around ten days indicated as being good sowing days. It so happened that during this time radish seeds from the same seed packet were sown on four different days. I noticed very early on that the seedlings displayed uneven growth patterns and as they developed, evolved quite distinct characteristics. Finding the cause of this development became a burning quest for me. So I then decided to carry out some daily sowing trials in the hope of solving the riddle.

The trials were set up on well cultivated (biodynamic) sandy soil with plenty of humus. It soon became apparent that different preceding crops altered the otherwise uniform results. So in order to have sufficient land available with the same preceding crop we had to take on additional garden areas. The soils were all treated with ripe compost that had received the biodynamic compost preparations. The horn manure preparation was only applied in spring and fall – a daily application for every sowing would have been impossible and might also have altered the plants' characteristics.

In the beginning a piece of land was cultivated and the seed bed prepared in advance. I then sowed a row of a certain number of seeds each day. After a while we found that while seeds sown on the first days had some variation, as time went on they became more and more uniform.

New issues were thus appearing that were not so easy to solve. A new bed had meanwhile been set up to undertake a further month of sowings and once again the same phenomenon occurred. After much heart searching we came to realize that it might have to do with the moment when the ground had been prepared. From then on we cultivated narrow strips and prepared the seed bed on the day we sowed the seeds. We soon saw considerable variation.

Further challenges then confronted us. A dry period occurred with a lot of hot weather. Traditional practice meant that the radish beds had to be irrigated. The result was that all the rows looked nearly identical. The only differences were in size due to the date of sowing, otherwise only the effects of watering could be seen. These experiences led us to decide that all future sowings would be left to the mercy of the weather. The only interventions would be hoeing and weeding. This was the basis upon which all future growing trials would be carried out.

I had discovered meanwhile that the development of a plant's root system proceeded in rhythmic way. This raised a new question: If the part of the plant which is above the earth is stimulated by the forces of the Full Moon, does it not follow that root growth is stimulated by the forces of the New Moon? This was not something readily visible in seedlings germinating in water.

I felt that the broad bean with its strong root would be a suitable plant to test. So I planted a few beans outside each day and kept them under observation. I thought after a few months that I had already discovered the rhythm – root growth always began on the day of the descending Moon node. Since the various lunar and planetary cycles were still a mystery for me, I was delighted by my discovery. The summer came and went and winter brought a pause to my investigations. The work was taken up again the following year. Now we found that this time root development began the day after the Moon node. This was very disappointing but it also made us determined to finally solve the problem. This cycle connected to the forming of roots had in the meantime unlocked new practical possibilities for the grower in terms of saving time when pricking out seedlings during the growing season.

The year 1956 came around and broad beans were again planted from early spring. The phenomena of the previous year repeated themselves. During the course of the fall the time difference increased further and by the following year root growth started a full three days after the Moon node. We then realized that making a link to the Moon node was a mistake.

Ascending and descending Moon

I felt it necessary then to study lunar and planetary rhythms more intensely. I soon discovered that when the first root observation had been made, the Moon was at its highest position in the zodiac and that this occurred one day before the Moon node. It remained on that day for a while and then in succeeding years was delayed by one, two or three days as the node gradually moved through the zodiac. The day in question was therefore determined by the sidereal Moon rhythm which in turn depends on the position of the Moon in the zodiac. At the time the node was positioned in the constellation of Gemini. This solved the mystery but further evidence was needed for it to be conclusive. Various types of cutting including runners from house plants and hyacinths, were then taken, put in water to root and observed over a long period. Parallel trials were also carried out on open ground with lettuce, cabbage, celery and annual flowers.

The effect of the descending Moon (that is, when the Moon's daily arcs become lower as it descends from the constellation of Gemini to Scorpio) was always confirmed and even strengthened when the descending forces of the day were used as well. Time and again we could observe how plants transplanted in the afternoon during the descending Moon needed to be watered only once at the moment of planting and would then grow on unhindered.

This process can be further improved if the plants are taken out of the ground in the morning while they are still subject to the ascending forces of the day, kept cool and moist over midday and then transplanted in the afternoon. Overnight the plants are then able to immediately orientate themselves to their new position and stand fresh and upright the next day. In comparison with plants of the same species remaining in the seed bed, there was better growth and vitality throughout the growing season. Plants of the same species transplanted during an ascending Moon period often struggle for days or even weeks. The outer leaves wilt and only with a great deal of effort is the plant able to renew itself out of its growing point. In practice such plants frequently require ongoing watering and a delay of several weeks can be detected in their overall development.

The Moon in the zodiac

Taking account of our ongoing experiences we continued with regular sowings of radish under the same parameters. A certain underlying principle could now be identified. However, we lacked the ability to read it. We found for example how the same type of plant developed over a period of two, three or sometimes four days. In one case there would be a strong root with very small leaves (Root type). In another there would be massive leaf development with relatively small roots (Leaf type). In the third case the plant developed a reddish stem and rapidly developed flowers which remained for a relatively long time but developed virtually no roots (Flower type). There was a fourth type which appeared to be in even more of a hurry. Forming roots seemed completely unnecessary, nor was much time spent on growing leaves. It was in a great hurry to produce flowers and then move on into producing seeds and reproducing (Fruit/Seed type).

A clear fourfold quality was revealed in the plant. This wasn't easy to find at first because we were convinced that the plant manifests itself in terms of the threefold organs of root, leaf and flower. The fourfold qualities that were found repeat themselves three times in the course of a month.

I suddenly realized that I might have found a basic law of life determining the way a plant is formed. Did those original breeders of our cultivated plants have an insight into this law? Gaining knowledge of this membering is an unimaginable reward for all these years of effort.

During the quiet days of winter I set about working through all these results in order to discover the origin of these formative processes. This led to the discovery that as the Moon passes from one constellation of the zodiac to the next, the plant type changes. I drew this pictorially and graphically and recalled to my amazement that I had seen a related differentiation of zodiac forces in a presentation given by Guenther Wachsmuth when he spoke about the etheric formative forces that stream out from the zodiac. These correlate in the following way:

Wachsmuth's observation		My plant observations
Life ether	Taurus	Root formation
Life ether	Virgo	Root formation
Life ether	Capricorn	Root formation
	Earth forces	
Chemical ether	Pisces	Leaf formation
Chemical ether	Cancer	Leaf formation
Chemical ether	Scorpio	Leaf formation
	Watery forces	
Light ether	Gemini	Flower formation
Light ether	Libra	Flower formation
Light ether	Aquarius	Flower formation
	Light forces	
Warmth ether	Aries	Fruit formation
Warmth ether	Leo	Fruit formation
Warmth ether	Sagittarius	Fruit formation
	Warmth forces	

The logical consequence is that plant growth is stimulated by one of the four etheric forces relating to the classical elements of Earth, Water, Air and Fire, through certain lunar forces depending on the Moon's position in the zodiac (as with my plant observations). The Moon takes on the role of mediator – at least in terms of the cultivated plant species (radishes) that have been observed – between the realm of the fixed stars and the elements of plant growth.

We then tried to understand what caused an upset to the processes of growth on certain days. We found that the Moon's perigee almost always has a negative effect and that the days of the Moon nodes are often unfavorable. They were found to cause blockages to growth. Other influences which override a particular plant type arise through planetary oppositions. The opposition of Venus and Jupiter, for example, encouraged a powerful flowering impulse.

Radishes from Leaf times

Trials with other plants

Many new questions arose, in particular, how do these forces affect other culti-vated plants – or is perhaps only the radish a suitable medium? The challenge now was to test 'typical' representative plants. During the following years tests were carried out on carrots, parsnips and salsify as Root crops; lettuce, spinach, corn salad, cress and, to a small extent, cabbages as Leaf plants; zinnias, snap-dragons and asters as Flowers; and beans, peas, cucumbers and to a small extent tomatoes as Fruiting plants.

Prepared plant compost was applied, and in spring and fall the horn manure preparation. Horn silica was not used, but since then we have found that when it is used intensively the quality of plant type can be enhanced.

The experiences made with radishes were confirmed with other annual plant species. It was often difficult to discover significant differences in the early growth stages of a plant but on reaching maturity the result always came out the same. Root vegetables that were sown when the Moon was in Taurus, Virgo and Capricorn developed the best roots as well as the most harmonious structure and flavorsome quality while the upper part of the plant remained relatively small. With Root vegetables sown at Leaf times there was vigorous leaf devel-opment and small roots whose keeping quality was moderate. Roots shrivelled quite early during storage unlike those sown at Root times which kept in the sand well into spring and had a somewhat salt-like structure. Root vegetables sown at Flower and Fruit times had very finely divided leaves while the roots were often stringy and sometimes even slightly woody. In the warm year of 1959 several of them even went to seed.

Plants which are valued primarily for their leaves such as spinach, lettuce, corn salad and cress (cabbage family) thrive best when sown with the Moon in Pisces, Cancer and Scorpio. These plants develop a mass of leaves and remain at

Radishes from Root times

the leaf stage without going rapidly to seed. If the plants are left too long they start to tiller – a sign that the reproductive power is retained in the leaf realm. This latter symptom is observed most strongly with sowings in Cancer and Scorpio. In this way it is possible to sow corn salad and spinach from spring through to late fall without fear of the plants bolting. Another small difference could be observed in that sowings made before St John's tide in Cancer and Scorpio and after St John's in Pisces, had the greater leaf mass.

In the case of snapdragons, asters and zinnias we made the following observations: Sowings made in Pisces, Cancer and Scorpio resulted in a large amount of leaf growth but flowering only occurred with great difficulty; one could say there was a certain 'flowering laziness'. Sown at Fruit times these plants rapidly produced a few tiny leaves, went quickly to flower and then formed seed. The most harmonious Flowering plants came from sowings made with the Moon in Gemini, Libra and Aquarius. Branching clumps developed with a plethora of flower buds and the plants remained in the flowering state for a long period. Flowers can be cut continuously for the plant is always producing new side shoots. There is a pronounced propensity for flowering. Flowers sown at Root times have a meagre growth above ground and little tendency for flowering.

In observing the Fruiting plants new questions emerged. We found that although the constellations of Aries and Sagittarius were particularly favorable for developing the fleshy fruits of cucumbers, tomatoes and beans, the harvest of ripe peas and beans was not so satisfactory. For a good crop of ripe peas and broad beans the Moon in Leo was found to be best. When cucumbers, beans and tomatoes were sown with the Moon in Leo the plants went through their young growth stage and very quickly formed fruits with large seeds. I made the provisional assumption that this was the result of other influences – which is why new trials were continually being set up. We had found a fourfoldness – was this now being put in question?

Sown at Leaf times cucumbers, tomatoes and beans responded with the now familiar phenomenon – a huge amount of leaf growth and low propensity to flower. We found time and again that at Flower times cucumbers produced a large number of male flowers but very few fruit-bearing female flowers. At these different times the plant was actually only able to reproduce its own kind which for the gardener of course, is not enough. While the above-mentioned plants produced very desirable fruits when sown with the Moon in Aries and Sagittarius, only very poor growth above the earth occurred when Root times were used. Symptoms of congestion were repeatedly visible and this often led to an attack of aphids.

We found that climbing beans sown during the descending Root constellation of Virgo produced a huge root mass, the roots went very deep, grew thick as turnips and consisted to a great extent of nitrogen nodules. Something similar was found to be the case with peas and broad beans. These were the best times for sowing green manure crops too – their value is found in the root mass.

Further observations of fruit crops sown in Leo, Sagittarius and Aries showed that a distinction can be made between fruit and seed crops – there is in effect a fivefoldness whereby Sagittarius and Aries are Fruit times and Leo a Seed time.

Other observations

While writing down my observations from the summer of 1962 I discovered a lecture given in another context by Rudolf Steiner on January 1, 1912, where he describes the fivefold plant and its therapeutic effect on the human being. He pointed to the healing effect of plant seeds on the human heart. Elsewhere he also showed that within the physical body the region of the heart has its origin in the constellation of Leo. It is therefore remarkable yet quite fitting to discover that a plant medicine for the heart is formed and influenced by the same constellation.

This discovery gave me the courage to make the results of my work more widely available to those active in biodynamic agriculture. I am of course aware that there is still much to be discovered. The perennial plants, cereals, trees and other things all need to be carefully observed and then other rhythms will come into play.

From the outset the work which has been described demanded a precise observation of atmospheric conditions, the weather and also the micro-climate which is influenced by the lay of the land, the way dew forms, the breathing of the earth, wind direction, retention of warmth, and so on. In the process it became clear that the longer term weather conditions are determined by the movement of the planets through the various regions of the zodiac. Interruptions and disturbances are the result of conjunctions and oppositions. Hence the earthquakes we have experienced in recent years have always occurred during Neptune quadratures (square aspect) while powerful thunderstorms are prominent during a Uranus quadrature.

The etheric qualities which can be observed in the micro-climate on the other hand, are strongly connected with the passage of the Moon through the zodiac. The formation of dew and mist at Water times is completely different to that at Warmth times. When the Moon is transiting Pisces, Cancer and Scorpio the sky is generally grey and overcast and if this is reinforced by longer term influences, then rain is to be expected at those times. If it is not supported by the moisture bringing influences of the planets then a layer of mist will at least bring more moisture than at other times. During periods of drought it is possible to retain this moisture using appropriate soil cultivations. During cold and damp periods we have always observed how the atmosphere at Warmth and Light times has quite a different quality than at Water or Earth times. It is possible in this way to bring more warmth and light into the soil and retain or release moisture depending on what the plants require.

Work on the crop after transplanting

In the early years we found that seed bed preparation on the day of sowing had a significant effect on the development of the plant. This led to the question, can the particular crop type be further supported by cultivating the soil at the corresponding times?

The results described above came from humus-containing sandy soils and later on from clay soil. Both soils had been under well-managed biodynamic cultivation.

During the course of the sowing trials an ever stronger response from the plants to these cosmic rhythms could be observed. Once these results had attained a certain level of accuracy they were passed on for other people to try. When checking these myself I came to realize that the response of strongly mineralized soils to these finer influences was minimal, while good humus-rich soils of whatever type were capable of responding strongly to these forces.

The question has often been put, what is a living soil? Here it became visible – a soil which as a result of applying well-prepared compost, possesses the power to absorb cosmic influences, transform them into positive formative processes and enable the plant itself to activate forces from the stars and thereby create the quality nutrition that human beings need.

Background to the Calendar

The zodiac

The **zodiac** is a group of twelve constellations of stars which the Sun, Moon and all the planets pass on their circuits. The Sun's annual path always takes exactly the same line, called **ecliptic.**

The angles between the Sun, Moon and planets are called **aspects.** In this calendar the most important is the 120° angle, or trine.

In the illustration below the outer circle shows the varying sizes of the visible **constellations** of the zodiac. The dates on this outer circle are the approximate dates on which the Sun enters the constellation (from year to year the actual date can change by one day because of leap years). The inner circle shows the divisions into equal sections of 30° corresponding to the **signs** used in astrology.

It is the *constellations,* not the signs, on which our observations are based, and which are used throughout this calendar.

The twelve constellations are grouped into four different types, each having three constellations at an angle of about 120°, or trine. About every nine days the Moon passes from one type – for instance Root – through the other types (Flower, Leaf and Fruit) back to the same type again.

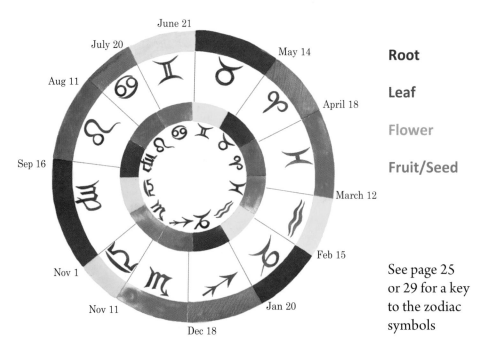

Root

Leaf

Flower

Fruit/Seed

See page 25 or 29 for a key to the zodiac symbols

22

What are oppositions, trines, conjunctions and nodes?

Oppositions ☍

A **geocentric** (Earth-centred) **opposition** occurs when for the observer on the Earth there are two planets opposite one another – 180° apart – in the heavens. You cannot see both planets at opposition – one will be above the horizon, the other below. They look at one another from opposite sides of the sky and their light interpenetrates. Their rays fall on to the Earth and stimulate in a beneficial way the seeds that are being sown in that moment. In our trials we have found that seeds sown at times of opposition resulted in a higher yield of top quality crops.

At times of opposition two zodiac constellations are also playing their part. If one planet is standing in a Warmth constellation, the second one will usually be in a Light constellation or vice versa. If one planet is in a Water constellation, the other will usually be in an Earth one. (As the constellations are not equally sized, the point opposite may not always be in the opposite constellation.)

With a heliocentric (Sun-centred) opposition the Sun is in the centre and the two planets placed 180° apart also gaze at each other but this time across the Sun. Their rays are also felt by the Earth and stimulate better plant growth. However, heliocentric oppositions are not shown or taken into account in the calendar.

Trines △ or ▲

The twelve constellations are grouped into four different types, each having three constellations at an angle of about 120°, or trine. About every nine days the Moon passes a similar region of forces.

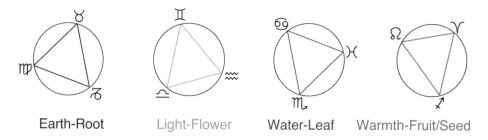

Earth-Root   Light-Flower   Water-Leaf   Warmth-Fruit/Seed

Trines occur when planets are 120° from one another. The two planets are then usually both standing in the same elemental configuration – Aries and Leo for example are both Warmth constellations. A Warmth trine means that the effects of these constellations will enhance fruit and seed formation in the plants sown at this time. If two planets are in trine position in Water, watery influences will be enhanced, which usually brings high rainfall. Plants sown at these times will yield more leaf than those at other times. Trine effects can change the way plants grow.

Sometimes when two planets are in a trine, they are in different types of constellations. These are shown as △ under planetary aspects, but have no effect on plant growth and are not shown as colored ▲ in the left-hand pages.

Conjunctions ☌

Conjunctions occur when two or more planets stand behind one another in space. It is then usually only the planet closest to the Earth that has any influence on plant growth. If this influence is stronger than that of the Moon, cosmic disturbances can occur that irritate the plant and cause checks in growth. This negative effect is increased further when the Moon or another planet stands directly in front of another – an occultation (•) or eclipse in the case of Sun and Moon. Sowing at these times will affect subsequent growth detrimentally and harm a plant's regenerative power.

Nodes ☊ and ☋

The Moon's and planets' paths vary slightly, sometimes above and sometimes below the ecliptic. The point at which their paths cross the ecliptic is called a node (☊ is the symbol for ascending node, and ☋ for the descending node). Nodes cause unfavorable times of varying lengths.

If a New Moon is at a node there is a solar eclipse, as the Moon is directly in front of the Sun, while a Full Moon at a node causes a lunar eclipse where the Earth's shadow falls on the Moon. If the Sun or Moon pass exactly in front of a planet, there is an occultation (•). If Mercury or Venus pass exactly in front of the Sun, this is a transit (other planets cannot pass in front of the Sun).

The effects of the Moon

In its 27-day orbit round the Earth, the Moon passes through the constellations of the zodiac and transmits forces to the Earth which affect the four elements: Earth, Light (Air), Water and Warmth (Fire). They in turn affect the four parts of the plant: the roots, the flower, the leaves and the fruit or seeds. The health and growth of a plant can therefore be stimulated by sowing, cultivating and harvesting it in tune with the cycles of the Moon.

These cosmic forces can also be harnessed in beekeeping. By opening and closing the bee 'skep' or box in rhythm with the Moon, the bees' activity is directly affected.

The table below summarizes the effects of the movement of the Moon through the twelve constellations on plants, bees and the weather.

The amount of time the Moon spends in any constellation varies between two and four days. However, this basic framework can be disrupted by planetary oppositions which override the normal tendencies; equally, it may be that trine positions (see pp. 23f) activate a different elemental force to the ones the Moon is transmitting. Times when the Moon's path or a planet's path intersects with the ecliptic (ascending ☊ or descending ☋ node; see page 24) are subject to mainly negative effects. These are intensified if there is an eclipse or occultation, in which case the nearer planet interrupts the influence of the distant one. Such times are unsuitable for sowing or harvesting.

Constellation	Sign	Element	Plant	Bees	Weather
Pisces, Fishes	♓	W Water	Leaf	Making honey	Damp
Aries, Ram	♈	H Warmth	Fruit	Gathering nectar	Warm/hot
Taurus, Bull	♉	E Earth	Root	Building comb	Cool/cold
Gemini, Twins	♊	L Light	Flower	Gathering pollen	Airy/bright
Cancer, Crab	♋	W Water	Leaf	Making honey	Damp
Leo, Lion	♌	H Warmth	Fruit	Gathering nectar	Warm/hot
Virgo, Virgin	♍	E Earth	Root	Building comb	Cool/cold
Libra, Scales	♎	L Light	Flower	Gathering pollen	Airy/bright
Scorpio, Scorpion	♏	W Water	Leaf	Making honey	Damp
Sagittarius, Archer	♐	H Warmth	Fruit	Gathering nectar	Warm/hot
Capricorn, Goat	♑	E Earth	Root	Building comb	Cool/cold
Aquarius, Waterman	♒	L Light	Flower	Gathering pollen	Airy/bright

Good Friday to Easter

Easter is a date set by astronomical events: it is determined by the Full Moon after the spring equinox. Our experience and trials over the last 40 years have shown that Good Friday and the Saturday are not good times for sowing or transplanting. Seeds sown on those days germinate poorly, plants transplanted on these days don't root properly and most don't survive. This negative effect on plant growth begins in the early morning of Good Friday and ends at sunrise on Easter Sunday, *local time*. This is why these days are marked as unfavorable in the calendar. Remember to apply *local time* for this effect, as the unfavorable time shown is for Easter Daylight Time.

Groupings of plants for sowing and harvesting

When we grow plants, different parts are cultivated for food. We can divide them into four groups.

Root crops at Root times

Radishes, swedes, sugar beet, beetroot, celeriac, carrot, scorzonera, etc., fall into the category of root plants. Potatoes and onions are included in this group too. Root times produce good yields and top storage quality for these crops. Green manure crops are best sown at Root times.

Leaf plants at Leaf times

The cabbage family, lettuce, spinach, lambs lettuce, endive, parsley, leafy herbs and fodder plants are categorized as leaf plants. Leaf times are suitable for sowing and tending these plants but not for harvesting and storage. For this (as well as harvesting of cabbage for sauerkraut) Fruit and Flower times are recommended.

Flower plants at Flower times

These times are favorable for sowing and tending all kinds of flower plants but also for cultivating and spraying 501 (a biodynamic preparation) on oil-bearing plants such as linseed, rape, sunflower, etc. Cut flowers have the strongest scent and remain fresh for longer if cut at Flower times, and the mother plant will provide many new side shoots. If flowers for drying are harvested at Flower times they retain the most vivid colors. If cut at other times they soon lose their color. Oil-bearing plants are best harvested at Flower times.

Fruit Plants at Fruit times

Plants that are cultivated for their fruit or seed belong to this category, including beans, peas, lentils, soya, maize, tomatoes, cucumber, pumpkin, zucchini, but also cereals for summer and winter crops. Sowing oil-bearing plants at Fruit times provides the best yields of seeds. The best time for extraction of oil later on is at Flower times. Leo times are particularly suitable to grow good seed. Fruit plants are best harvested at Fruit times. They store well and their seeds provide good plants for next year. When storing fruit, also remember to choose the time of the ascending Moon.

There is always uncertainty as to which category some plants belong (see list opposite). Onions and beetroot provide a similar yield when sown at Root and Leaf times, but the keeping quality is best from Root times. Broccoli is more beautiful and firmer when sown at Flower times.

Types of crop

Flower plants

artichoke
broccoli
flower bulbs
flowering ornamental shrubs
flowers
flowery herbs
rose
sunflower

Leaf plants

asparagus
Brussels sprouts
cabbage
cauliflower
celery
chard
chicory (endive)
Chinese cabbage (pe-tsai)
corn salad (lamb's lettuce)
crisphead (iceberg) lettuce
curly kale (green cabbage)
endive (chicory)
finocchio (Florence fennel)
green cabbage (curly kale)
iceberg (crisphead) lettuce
kohlrabi
lamb's lettuce (corn salad)
leaf herbs
leek
lettuce
pe-tsai (Chinese cabbage)
red cabbage
rhubarb
shallots
spinach

Root plants

beetroot
black (Spanish) salsify
carrot
celeriac
garlic
green manure crops
horseradish
Jerusalem artichoke
parsnip
potato
radish
red radish
root tubers
Spanish (black) salsify

Fruit plants

aubergine (eggplant)
bush bean
courgette (zucchini)
cucumber
eggplant (aubergine)
grains
lentil
maize
melon
paprika
pea
pumpkin (squash)
runner bean
soya
squash (pumpkin)
tomato
zucchini (courgette)

Explanations of the Calendar Pages

Next to the date is the constellation in which the Moon is positioned (or constellations, with time of entry into the new one). This is the astronomical constellation, not the astrological sign (see page 22). The next column shows solar and lunar events.

A further column shows which element is dominant on that day (this is also useful for beekeepers). Note H is used for warmth (heat). If there is a change during the day, both elements are mentioned.

The vertical green color band ■ shows Northern Transplanting Time (see next page).

The next column shows in color the part of the plant which will be enhanced by sowing or cultivation on that day. Numbers indicate times of day. The column of special events in nature (like storms, thunderstorms, etc.) is being discontinued as weather patterns seem to be changing, and earthquakes and volcanic tendencies are of little help for gardening or agricultural activities.

When parts of the plant are indicated that do not correspond to the Moon's position in the zodiac (often it is more than one part on the same day), it is not a misprint, but takes account of other cosmic aspects (for instance, the Moon's Apogee or trines) which overrule the Moon-zodiac pattern and have an effect on a different part of the plant.

Unfavorable times are marked thus ▬. These are caused by eclipses, nodal points of the Moon or the planets or other aspects with a negative influence; they are not elaborated upon in the calendar. If one has to sow at unfavorable times for practical reasons, one can choose favorable times for hoeing, which will improve the plant.

The *position of the planets* in the zodiac is shown in the box below, with the date of entry into a new constellation. R indicates the planet is moving retrograde (with the date when retrograde begins), D indicates the date when it moves in direct motion again.

On the opposite calendar page astronomical aspects are indicated. Those visible to the naked eye are shown in **bold** type. Visible conjunctions (particularly Mercury's) are not always visible from all parts of the Earth, because the planets may be too close to the horizon in highter latitudes.

Astronomical symbols

Constellations		Planets		Aspects			
♓	Pisces	☉	Sun	♌	Ascending node	E	Earth element
♈	Aries	☽, ☾	Moon	☋	Descending node	L	Light/air element
♉	Taurus	☿	Mercury	⌢	Highest Moon	W	Water element
♊	Gemini	♀	Venus	⌣	Lowest Moon	H	Warmth (heat)
♋	Cancer	♂	Mars	**Pg**	Perigee		
♌	Leo	♃	Jupiter	**Ag**	Apogee		
♍	Virgo	♄	Saturn	☍	Opposition		Northern Trans-
♎	Libra	♅	Uranus	☌	Conjunction		planting Time
♏	Scorpio	♆	Neptune	☌•	Eclipse/occultation		
♐	Sagittarius	♇	Pluto	☌•	Lunar eclipse		
♑	Capricorn	○	Full Moon	△	Trine (or ▲)		
♒	Aquarius	●	New Moon				

Transplanting times

From midwinter through to midsummer the Sun rises earlier and sets later each day while its path across the sky ascends higher and higher. From midsummer until midwinter this is reversed, the days get shorter and the midday Sun shines from an ever lower point in the sky. This annual ascending and descending of the Sun creates our seasons. As it ascends and descends during the course of the year, the Sun is slowly moving (from an Earth-centred point of view) through each of the twelve constellations of the zodiac in turn. On average it shines for one month from each constellation.

In the northern hemisphere the winter solstice occurs when the Sun is in the constellation of Sagittarius and the summer solstice when it is in Gemini. At any point from Sagittarius to Gemini the Sun is ascending, while from Gemini to Sagittarius it is descending. In the southern hemisphere this is reversed.

The Moon (and all the planets) follow approximately the same path as the Sun around the zodiac but instead of a year, the Moon takes only about 27½ days to complete one cycle, shining from each constellation in turn for a period of two to three days. This means that the Moon will ascend for about fourteen days and then descend.

It is important to distinguish the journey of the Moon through the zodiac

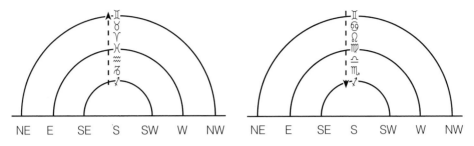

NE E SE S SW W NW   NE E SE S SW W NW

Northern hemisphere ascending Moon (left) and descending Moon (right): Transplanting Time

(siderial rhythm) from the waxing and waning (synodic) cycle: in any given constellation there may be a waxing, waning, full, quarter, sickle or gibbous Moon. As it moves through the zodiac the Moon, like the Sun, is ascending (in the northern hemisphere) when it is in the constellations from Sagittarius to Gemini and descending from Gemini to Sagittarius. In the southern hemisphere it is ascending from Gemini to Sagittarius and descending from Sagittarius to Gemini.

When the Moon is ascending, plant sap rises more strongly. The upper part of the plant fills with sap and vitality. This is a good time for cutting scions (for grafting). Fruit harvested during this period remains fresh for longer when stored.

When the Moon is descending, plants take root readily and connect well with their new location. This period is referred to as the **Transplanting Time.** Moving plants from one location to another is called transplanting. This is the case when young plants are moved from the seedbed into their final growing position but also when the gardener wishes to strengthen the root development of young fruit trees, shrubs or pot plants by frequently re-potting them. Sap movement is slower during the descending Moon. This is why it is a good time for trimming hedges, pruning trees and felling timber as well as applying compost to meadows, pastures and orchards.

Note that sowing is the moment when a seed is put into the soil; either the ascending or descending period can be used. It then needs time to germinate and grow. This is different from *transplanting*, which is best done during the descending Moon. These times are given in the calendar.

Northern Transplanting Times refer to the northern hemisphere, and **Southern Transplanting Times** refer to the southern hemisphere. All other constellations and planetary aspects are equally valid in both hemispheres.

Converting to local time

Times given are Eastern Standard Time (EST), or from March 14 to Nov 6 Eastern Daylight Saving Time (EDT), with ᵃ or ₚ after the time for am and pm.

Noon is 12_p and midnight is 12_a; the context shows whether midnight at the beginning of the day or at the end is meant; where ambiguous (as for planetary aspects) the time has been adjusted by an hour for clarity.

For different time zones adjust as follows:

Newfoundland Standard Time: add $1\frac{1}{2}^h$
Atlantic Standard Time: add 1^h
Eastern Standard Time: do not adjust
Central Standard Time: subtract 1^h
For Saskatchewan subtract 1^h, but subtract 2^h from March 8 to Oct 31 (no DST)
Mountain Standard Time: subtract 2^h
For Arizona subtract 2^h, but subtract 3^h from March 8 to Oct 31 (no DST)
Pacific Standard Time: subtract 3^h
Alaska Standard Time: subtract 4^h
Hawaii Standard Time: subtract 5^h, but subtract 6^h from March 8 to Oct 31 (no DST)

For Central & South America adjust as follows:

Argentina: add 2^h, but add 1^h from March 14 to Nov 6 (no DST)
Brazil (Brasilia): add 2^h, but add 1^h from March 14 to Nov 6 (no DST)
Chile: Jan 1 to March 13 add 2^h; March 14 to April 3 add 1^h; April 4 to Sep 4 do not adjust; Sep 5 to Nov 6 add 1^h; from Nov 7 add 2^h.
Columbia, Peru: do not adjust, but subtract 1^h from March 14 to Nov 6 (no DST).
Mexico (mostly CST): subtract 1^h, but from March 14 to April 3, and from Oct 31 to Nov 6, subtract 2^h

For other countries use *The Maria Thun Biodynamic Calendar* from Floris Books which carries all times in GMT, making it easier to convert to another country's local and daylight saving time.

January 2021

All times in EST

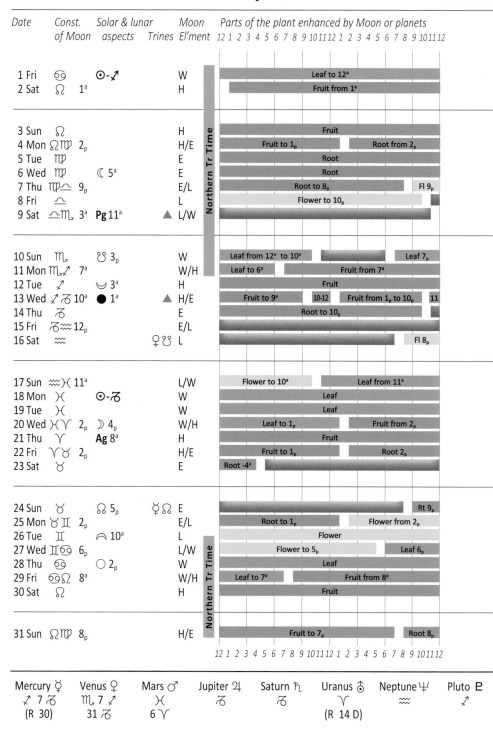

Date	Const. of Moon	Solar & lunar aspects	Trines	Moon El'ment	Parts of the plant enhanced by Moon or planets

1 Fri ♋ ☉-♐ W — Leaf to 12ª

2 Sat ♌ 1ª H — Fruit from 1ª

3 Sun ♌ H — Fruit

4 Mon ♌♍ 2ₚ H/E — Fruit to 1ₚ / Root from 2ₚ

5 Tue ♍ E — Root

6 Wed ♍ ☾ 5ª E — Root

7 Thu ♍♎ 9ₚ E/L — Root to 8ₚ / Fl 9ₚ

8 Fri ♎ L — Flower to 10ₚ

9 Sat ♎♏ 3ª Pg 11ª ▲ L/W

10 Sun ♏ ☍ 3ₚ W — Leaf from 12ª to 10ª / Leaf 7ₚ

11 Mon ♏♐ 7ª W/H — Leaf to 6ª / Fruit from 7ª

12 Tue ♐ ☌ 3ª H — Fruit

13 Wed ♐♑ 10ª ● 1ª ▲ H/E — Fruit to 9ª / 10-12 / Fruit from 1ₚ to 10ₚ / 11

14 Thu ♑ E — Root to 10ₚ

15 Fri ♑♒ 12ₚ E/L

16 Sat ♒ ♀☍ L — Fl 8ₚ

17 Sun ♒♓ 11ª L/W — Flower to 10ª / Leaf from 11ª

18 Mon ♓ ☉-♑ W — Leaf

19 Tue ♓ W — Leaf

20 Wed ♓♈ 2ₚ ☽ 4ₚ W/H — Leaf to 1ₚ / Fruit from 2ₚ

21 Thu ♈ Ag 8ª H — Fruit

22 Fri ♈♉ 2ₚ H/E — Fruit to 1ₚ / Root 2ₚ

23 Sat ♉ E — Root -4ª

24 Sun ♉ ♌ 5ₚ ☿♌ E — Rt 9ₚ

25 Mon ♉♊ 2ₚ E/L — Root to 1ₚ / Flower from 2ₚ

26 Tue ♊ ⌒ 10ª L — Flower

27 Wed ♊♋ 6ₚ L/W — Flower to 5ₚ / Leaf 6ₚ

28 Thu ♋ ○ 2ₚ W — Leaf

29 Fri ♋♌ 8ª W/H — Leaf to 7ª / Fruit from 8ª

30 Sat ♌ H — Fruit

31 Sun ♌♍ 8ₚ H/E — Fruit to 7ₚ / Root 8ₚ

Mercury ☿	Venus ♀	Mars ♂	Jupiter ♃	Saturn ♄	Uranus ♅	Neptune ♆	Pluto ♇
♐ 7 ♑	♏ 7 ♐	♓	♑	♑	♈	♒	♐
(R 30)	31 ♑	6 ♈			(R 14 D)		

NB: All zodiac symbols refer to astronomical constellations, not astrological signs (see p. 24)

Planetary aspects
(**Bold** = *visible to naked eye*)

Day	Aspects
1	
2	
3	
4	☽☍♆ 5ᵃ ☿☌♇ 8ₚ
5	
6	
7	☽☍♂ 4ᵃ ☽☍☋ 3ₚ
8	
9	♀△♂ 11ᵃ ☿☌♄ 10ₚ
10	
11	☿☌♃ 12ₚ **☽☌♀ 3ₚ**
12	
13	☽☌♇ 2ᵃ ☽☌♄ 5ₚ ♀△☋ 7ₚ **☽☌♃ 10ₚ**
14	**☽☌☿ 4ᵃ** ☉☌♇ 9ᵃ
15	
16	♀☋ 7ᵃ
17	☽☌♆ 5ᵃ
18	
19	
20	♂☌☋ 4ₚ
21	☽☌☋ 4ᵃ **☽☌♂ 4ᵃ**
22	
23	☉☌♄ 10ₚ
24	☿☊ 5ᵃ
25	
26	
27	☽☍♀ 11ᵃ ☽☍♇ 1ₚ
28	☽☍♄ 7ᵃ ♀☌♇ 11ᵃ **☽☍♃ 3ₚ** ☉☌♃ 9ₚ
29	**☽☍☿ 9ₚ**
30	
31	**☽☍♆ 12ₚ**

Planet (naked eye) visibility
Evening: Mercury (from 12th), Mars, Jupiter (to 16th), Saturn (to 9th)
All night: -
Morning: Venus (to 31st)

January 2021

With Uranus and Pluto in Warmth constellations, joined by Venus and Mars (from Jan 6–7), together with two Warmth trines, the start of the year is likely to be mild. After the first week, when Mercury joins Jupiter and Saturn in Capricorn, it may become a little cooler.

Northern Transplanting Time
Dec 30 to Jan 12 1ᵃ and
Jan 26 12ₚ to Feb 8
Southern Transplanting Time
Jan 12 5ᵃ to Jan 26 8ᵃ

The transplanting time is a good time for **pruning fruit trees, vines and hedges.** Fruit and Flower times are preferred for this work. Avoid unfavorable times.

When **milk processing** it is best to avoid unfavorable times. This applies to both butter and cheese making. Milk which has been produced at Warmth/Fruit times yields the highest butterfat content. This is also the case on days with a tendency for thunderstorms. Times of moon perigee (**Pg**) are almost always unfavorable for milk processing and even yoghurt will not turn out well. Starter cultures from such days decay rapidly and it is advisable to produce double the amount the day before. Milk loves Light and Warmth times best of all. Water times are unsuitable.

▬▬▬ *Unfavorable time*

Feb

February 2021

All times in EST

Date	Const. of Moon	Solar & lunar aspects	Trines	Moon El'ment	Parts of the plant enhanced by Moon or planets
1 Mon	♍	☉-♑		E	Root
2 Tue	♍			E	Root to 2_p
3 Wed	♍	**Pg** 3_p		E	Root from 4_p
4 Thu	♍♎ 2^a	☾ 1_p		E/L	1^a; Flower from 2^a
5 Fri	♎♏ 9^a			L/W	Flower to 8^a; Leaf from 9^a
6 Sat	♏	☋ 8_p		W	Leaf to 3_p
7 Sun	♏♐ 2_p			W/H	Leaf from 12^a to 1_p; Fruit from 2_p
8 Mon	♐	◡ 11^a		H	Fruit
9 Tue	♐♑ 7_p			H/E	Fruit to 6_p; Root 7_p
10 Wed	♑			E	Root
11 Thu	♑♒ 9_p	● 2_p		E/L	Root to 8_p; Fl 9_p
12 Fri	♒			L	Flower
13 Sat	♒♓ 7_p			L/W	Flower to 6_p; Leaf 7_p
14 Sun	♓	☉-♒		W	Leaf
15 Mon	♓			W	Leaf
16 Tue	♓♈ 10_p			W/H	Leaf to 9_p; 10_p
17 Wed	♈			H	Fruit
18 Thu	♈♉ 11_p	**Ag** 5^a		H/E	Fruit to 10_p; 11
19 Fri	♉	☽ 2_p		E	Root
20 Sat	♉	☊ 9_p		E	Root to 4_p
21 Sun	♉♊ 11_p			E/L	Root from 1^a to 10_p; 11
22 Mon	♊	◠ 7_p		L	Flower
23 Tue	♊			L	Flower
24 Wed	♊♋ 4^a			L/W	Fl -3^a; Leaf from 4^a
25 Thu	♋♌ 5_p			W/H	Leaf to 4_p; Fruit from 5_p
26 Fri	♌			H	Fruit
27 Sat	♌	○ 3^a		H	Fruit
28 Sun	♌♍ 4^a			H/E	Fr -3^a; Root from 4^a

Northern Tr Time (1–7 Feb; 22–28 Feb)

Hours: 12 1 2 3 4 5 6 7 8 9 10 11 12 1 2 3 4 5 6 7 8 9 10 11 12

Mercury ☿	Venus ♀	Mars ♂	Jupiter ♃	Saturn ♄	Uranus ⛢	Neptune ♆	Pluto ♇
♑	♑	♈	♑	♑	♈	♒	♐
(R 20 D)	22 ♒	20 ♉					

NB: All zodiac symbols refer to astronomical constellations, not astrological signs (see p. 24)

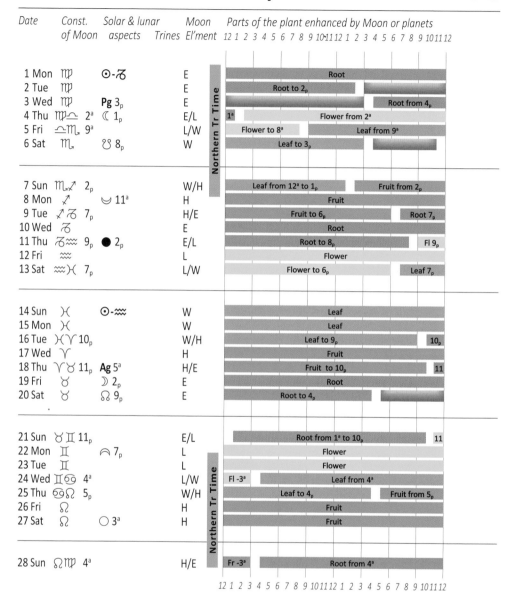

Planetary aspects
(**Bold** = *visible to naked eye*)

February 2021

1
2
3 ☽ ☍ ♁ 9$_p$
4 ☽ ☍ ♂ 10^a
5
6 ♀ ☌ ♄ 2^a

7
8 ☉ ☌ ☿ 9^a
9 ☽ ☌ ♇ 12$_p$
10 ☽ ☌ ♄ 8^a ☽ ☌ ♀ 5$_p$ ☽ ☌ ♃ 6$_p$
11 ☽ ☌ ☿ 2^a ♀ ☌ ♃ 10^a
12
13 ☿ ☌ ♀ 3^a ☽ ☌ ♆ 3$_p$

14 ☿ ☌ ♃ 5$_p$
15
16
17 ☽ ☌ ♁ 1$_p$
18 ☽ ☌ ♂ 8$_p$
19
20

21
22
23
24 ☽ ☍ ♇ 1^a ♂ △ ♇ 9$_p$ ☽ ☍ ♄ 10$_p$
25 ☽ ☍ ☿ 5^a ☽ ☍ ♃ 12$_p$
26 ☽ ☍ ♀ 3$_p$
27 ☽ ☌ ♆ 10$_p$

28

Mercury, Venus, Jupiter and Saturn together with the Sun in the first half of the month are all in Capricorn bringing cold. Mars (until Feb 21), Uranus and Pluto are in Warmth constellations which may counteract this. Only Neptune mediates some Light qualities.

Northern Transplanting Time
Jan 26 to Feb 8 9^a and
Feb 22 9$_p$ to March 7
Southern Transplanting Time
Feb 8 1$_p$ to Feb 22 5$_p$

Vines, fruit trees and shrubs can be pruned during Transplanting Time selecting Flower and Fruit times in preference. Avoid unfavorable times.

Best times for taking **willow cuttings for hedges and fences:** At Flower times outside Transplanting Time. In warm areas at Flower times during Transplanting Time to avoid too strong a sap current.

Planet (naked eye) visibility
Evening: Mercury (to 2nd), Mars
All night: -
Morning: -

■■■■ *Unfavorable time* 35

March (vertical side tab)

Date	Const. of Moon	Solar & lunar aspects	Moon Trines	El'ment	Parts of the plant enhanced by Moon or planets

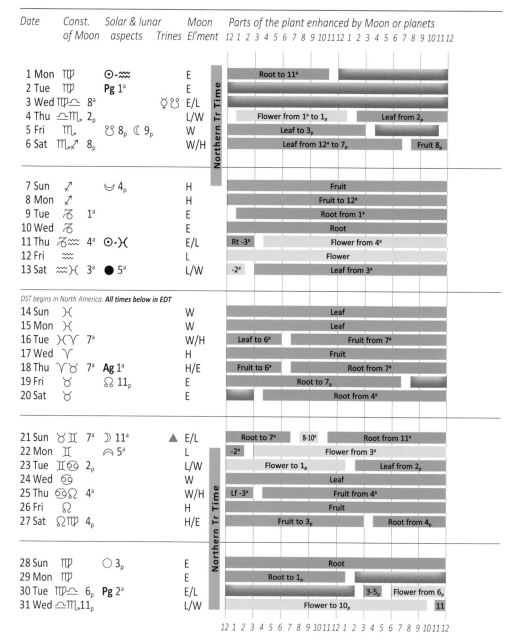

Column scale (top and bottom): 12 1 2 3 4 5 6 7 8 9 10 11 12 1 2 3 4 5 6 7 8 9 10 11 12

Northern Tr Time (vertical label)

1 Mon ♍	☉-♒	E	Root to 11ᵃ
2 Tue ♍	Pg 1ᵃ	E	
3 Wed ♍♎ 8ᵃ	☿ ☍	E/L	
4 Thu ♎♏ 2ₚ		L/W	Flower from 1ᵃ to 1ₚ — Leaf from 2ₚ
5 Fri ♏	☍ 8ₚ ☾ 9ₚ	W	Leaf to 3ₚ
6 Sat ♏♐ 8ₚ		W/H	Leaf from 12ᵃ to 7ₚ — Fruit 8ₚ

7 Sun ♐	☽ 4ₚ	H	Fruit
8 Mon ♐		H	Fruit to 12ᵃ
9 Tue ♑ 1ᵃ		E	Root from 1ᵃ
10 Wed ♑		E	Root
11 Thu ♑♒ 4ᵃ	☉-♓	E/L	Rt -3ᵃ — Flower from 4ᵃ
12 Fri ♒		L	Flower
13 Sat ♒♓ 3ᵃ	● 5ᵃ	L/W	-2ᵃ — Leaf from 3ᵃ

*DST begins in North America. **All times below in EDT***

14 Sun ♓		W	Leaf
15 Mon ♓		W	Leaf
16 Tue ♓♈ 7ᵃ		W/H	Leaf to 6ᵃ — Fruit from 7ᵃ
17 Wed ♈		H	Fruit
18 Thu ♈♉ 7ᵃ	Ag 1ᵃ	H/E	Fruit to 6ᵃ — Root from 7ᵃ
19 Fri ♉	☍ 11ₚ	E	Root to 7ₚ
20 Sat ♉		E	Root from 4ᵃ

21 Sun ♉♊ 7ᵃ	☽ 11ᵃ	E/L	Root to 7ᵃ — 8-10ᵃ — Root from 11ᵃ
22 Mon ♊	☌ 5ᵃ	L	-2ᵃ — Flower from 3ᵃ
23 Tue ♊♋ 2ₚ		L/W	Flower to 1ₚ — Leaf from 2ₚ
24 Wed ♋		W	Leaf
25 Thu ♋♌ 4ᵃ		W/H	Lf -3ᵃ — Fruit from 4ᵃ
26 Fri ♌		H	Fruit
27 Sat ♌♍ 4ₚ		H/E	Fruit to 3ₚ — Root from 4ₚ

Northern Tr Time (vertical label)

28 Sun ♍	○ 3ₚ	E	Root
29 Mon ♍		E	Root to 1ₚ
30 Tue ♍♎ 6ₚ	Pg 2ᵃ	E/L	3-5ₚ — Flower from 6ₚ
31 Wed ♎♏ 11ₚ		L/W	Flower to 10ₚ — 11

Column scale (bottom): 12 1 2 3 4 5 6 7 8 9 10 11 12 1 2 3 4 5 6 7 8 9 10 11 12

Mercury ☿	Venus ♀	Mars ♂	Jupiter ♃	Saturn ♄	Uranus ♅	Neptune ♆	Pluto ♇
♑ 13 ♒	♒	♉	♑	♑	♈	♒	♐
30 ♓	14 ♓						

NB: All zodiac symbols refer to astronomical constellations, not astrological signs (see p. 24)

Planetary aspects
*(**Bold** = visible to naked eye)*

March 2021

1
2
3 ☾☌☍☋ 4ᵃ ☿☍☊ 12ₚ
4 ☾☌♂ 7ₚ ☿☌♃ 10ₚ
5
6

7
8 ☾☌♇ 8ₚ
9 **☾☌♄ 8ₚ**
10 **☾☌♃ 1ₚ** ☉☌♆ 7ₚ ☾☌☿ 11ₚ
11
12 ☾☌♀ 11ₚ
13 ☾☌♆ 1ᵃ ♀☌♆ 11ₚ

14
15
16
17 ☽☌☋ 1ᵃ
18
19 ☽☌♂ 2ₚ
20

21 ♂△♄ 11ₚ
22
23 ☽☍♇ 11ᵃ
24 ☽☍♄ 2ₚ
25 ☽☍♃ 9ᵃ
26 ☉☌♀ 4ᵃ
27 ☽☍☿ 4ᵃ ☽☍♆ 11ᵃ

28 ☾☍♀ 4ₚ
29 ☿☌♆ 11ₚ
30 ☾☍☋ 4ₚ
31

In the first half Venus and in the second half of the month Mercury are in Aquarius, bringing Light, aided by Neptune. Mars, Jupiter and Saturn are in Earth constellations bringing cooler influences that are counteracted to some extent by Uranus and Pluto in Warmth constellations.

Northern Transplanting Time
Feb 22 to March 7 2ₚ and
March 22 7ᵃ to April 3
Southern Transplanting Time
March 7 6ₚ to March 22 3ᵃ

Willow cuttings for **pollen production** are best cut from March 22 7ᵃ to March 23 1ₚ; and for **honey flow** from March 25 4ᵃ to March 27 3ᵃ. Avoid unfavorable times.

Cuttings for grafting: Cut outside Transplanting Time during ascending Moon – always choosing times (Fruit, Leaf, etc.) according to the part of plant to be enhanced.

Control slugs: March 23 2ₚ to March 25 3ᵃ.

Biodynamic preparations
Pick dandelion in March or April in the mornings during Flower times. The flowers should not be quite open in the centre. Dry them on paper in the shade, not in bright sunlight. Once dried they can be stored until suitably encased and buried in the ground.

Planet (naked eye) visibility
Evening: Mars
All night:
Morning: Jupiter (from 1st), Saturn (from 4th)

Unfavorable time

April 2021

April

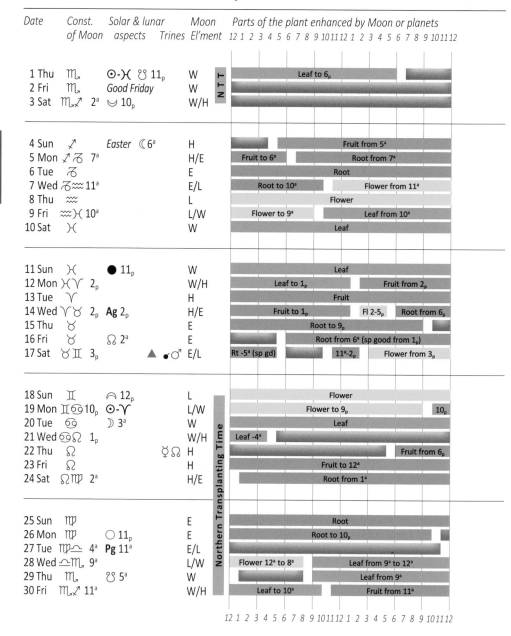

Date	Const. of Moon	Solar & lunar aspects	Moon Trines	El'ment	Parts of the plant enhanced by Moon or planets
1 Thu	♏	☉-♓ ☋ 11ₚ		W	Leaf to 6ₚ
2 Fri	♏	*Good Friday*		W	
3 Sat	♏♐ 2ᵃ	☍ 10ₚ		W/H	
4 Sun	♐	*Easter* ☾ 6ᵃ		H	Fruit from 5ᵃ
5 Mon	♐♑ 7ᵃ			H/E	Fruit to 6ᵃ / Root from 7ᵃ
6 Tue	♑			E	Root
7 Wed	♑♒ 11ᵃ			E/L	Root to 10ᵃ / Flower from 11ᵃ
8 Thu	♒			L	Flower
9 Fri	♒♓ 10ᵃ			L/W	Flower to 9ᵃ / Leaf from 10ᵃ
10 Sat	♓			W	Leaf
11 Sun	♓	● 11ₚ		W	Leaf
12 Mon	♓♈ 2ₚ			W/H	Leaf to 1ₚ / Fruit from 2ₚ
13 Tue	♈			H	Fruit
14 Wed	♈♉ 2ₚ **Ag** 2ₚ			H/E	Fruit to 1ₚ / Fl 2-5ₚ / Root from 6ₚ
15 Thu	♉			E	Root to 9ᵃ
16 Fri	♉	♌ 2ᵃ		E	Root from 6ᵃ (sp good from 1ₚ)
17 Sat	♉♊ 3ₚ	▲ ●♂	E/L	Rt -5ᵃ (sp gd) / 11ᵃ-2ₚ / Flower from 3ₚ
18 Sun	♊	⚹ 12ₚ		L	Flower
19 Mon	♊♋ 10ₚ	☉-♈		L/W	Flower to 9ₚ / 10ₚ
20 Tue	♋	☽ 3ᵃ		W	Leaf
21 Wed	♋♌ 1ₚ			W/H	Leaf -4ᵃ
22 Thu	♌	☿ ♌		H	Fruit from 6ₚ
23 Fri	♌			H	Fruit to 12ᵃ
24 Sat	♌♍ 2ᵃ			H/E	Root from 1ᵃ
25 Sun	♍			E	Root
26 Mon	♍	○ 11ₚ		E	Root to 10ₚ
27 Tue	♍♎ 4ᵃ **Pg** 11ᵃ			E/L	Flower 12ᵃ to 8ᵃ
28 Wed	♎♏ 9ᵃ			L/W	Leaf from 9ᵃ to 12ᵃ
29 Thu	♏	☋ 5ᵃ		W	Leaf from 9ᵃ
30 Fri	♏♐ 11ᵃ			W/H	Leaf to 10ᵃ / Fruit from 11ᵃ

Northern Transplanting Time

12 1 2 3 4 5 6 7 8 9 10 11 12 1 2 3 4 5 6 7 8 9 10 11 12

Mercury ☿	Venus ♀	Mars ♂	Jupiter ♃	Saturn ♄	Uranus ♅	Neptune ♆	Pluto ♇
♓ 19 ♈	♓	♉	♑	♑	♈	♒	♐
30 ♉	14 ♈	22 ♊	19 ♒			7 ♓	(27 R)

NB: All zodiac symbols refer to astronomical constellations, not astrological signs (see p. 24)

Planetary aspects

*(**Bold** = visible to naked eye)*

1		
2	☽☍♂ 7a	
3		
4		
5	☽☌♇ 3a	
6	**☽☌♄ 7a**	
7	**☽☌♃ 6a**	
8		
9	☽☌♆ 10a	
10		
11	☽☌☿ 5a	
12	☽☌♀ 8a	
13	☽☌⊕ 9a	
14		
15		
16		
17	♂△♃ 1a	☽•♂ 8a
18	☉☌☿ 10p	
19	☽☍♇ 8p	
20		
21	☽☍♄ 2a	
22	☽☍♃ 4a ☿☊ 5a ♀☌⊕ 9p	
23	☽☍♆ 11p	
24	☿☌⊕ 3a	
25	☿☌♀ 6p	
26		
27	☽☌⊕ 5a ☽☍♀ 2p ☽☍☿ 4p	
28		
29		
30	☉☌⊕ 4p ☽☍♂ 8p	

April 2021

In the first half of the month Mercury and Venus in Pisces, joined by Neptune on April 7, may bring precipitation. Until March 19–22 Mars, Jupiter and Saturn in Earth constellations bring cold, to some extent balanced by Uranus and Pluto in Warmth constellations.

Northern Transplanting Time
March 22 to April 3 8p and
April 18 2p to May 1
Southern Transplanting Time
April 4 1a to April 18 10a

The **soil warms up** on April 19.

Grafting of fruiting shrubs at Fruit times outside transplanting times.
Grafting of flowering shrubs at Flower times outside transplanting times.

Control
Slugs from April 19 10p to April 21 12p.
Clothes and wax moths from April 9 10a to April 12 1p.

Biodynamic preparations
The preparations can be taken out of the ground after April 28 avoiding unfavorable times (best at Fruit or Flower times). Preparations put into the ground after Sep 15, 2020, should wait to the end of May.

Planet (naked eye) visibility
Evening: Mercury (from 29th), Venus (from 20th), Mars
All night:
Morning: Jupiter, Saturn

Unfavorable time

May 2021

All times in EDT

Date	Const. of Moon	Solar & lunar aspects	Trines	Moon El'ment	Parts of the plant enhanced by Moon or planets
1 Sat	♐	☉-♈ ◡ 6^a		H	Fruit
2 Sun	♐♑ 2_p			H/E	Fruit to 1_p; Root from 2_p
3 Mon	♑	☾ 4_p		E	Root
4 Tue	♑♒ 4_p			E/L	Root to 3_p; Flower from 4_p
5 Wed	♒			L	Flower
6 Thu	♒♓ 4_p			L/W	Flower to 3_p; Leaf from 4_p
7 Fri	♓			W	Leaf
8 Sat	♓			W	Lf -3^a
9 Sun	♓♈ 8_p		♀☊	W/H	
10 Mon	♈			H	Fruit from 1^a
11 Tue	♈♉ 8_p	● 3_p **Ag** 6_p		H/E	Fruit to 8_p; Rt 9_p
12 Wed	♉		▲ ☌♀	E	Root to 3_p (sp good from 9^a); Rt 9_p
13 Thu	♉	*Ascension* ☊ 7^a		E	-2^a; Root from 11^a
14 Fri	♉♊ 9_p	☉-♉		E/L	Root to 8_p; Fl 9_p
15 Sat	♊	⌒ 6_p		L	Flower
16 Sun	♊			L	Flower
17 Mon	♊♋ 5^a			L/W	Fl to 4^a; Leaf from 5^a
18 Tue	♋♌ 9_p			W/H	Leaf to 8_p; Fr 9_p
19 Wed	♌	☽ 3_p	▲	H	Fruit to 2_p; Root from 3_p
20 Thu	♌			H	1^a; Fruit from 2^a
21 Fri	♌♍ 11^a			H/E	Fruit to 10^a; Root from 11^a
22 Sat	♍			E	Root
23 Sun	♍	*Pentecost*		E	Root
24 Mon	♍♎ 3_p			E/L	Root to 2_p; Flower from 3_p
25 Tue	♎♏ 8_p	**Pg** 10_p		L/W	Flower to 9^a
26 Wed	♏	☉☍○ 7^a ☋4_p		W	Leaf 8_p
27 Thu	♏♐ 9_p			W/H	Leaf to 8_p; Fr 9_p
28 Fri	♐	◡ 3_p		H	Fruit
29 Sat	♐♑ 10_p			H/E	Fruit to 11^a
30 Sun	♑		☿☋	E	
31 Mon	♑			E	Root from 1^a to 11_p

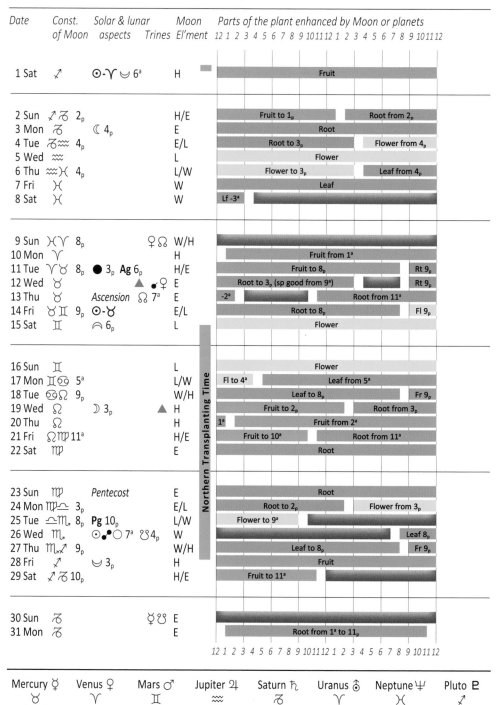

Northern Transplanting Time

12 1 2 3 4 5 6 7 8 9 10 11 12 1 2 3 4 5 6 7 8 9 10 11 12

Mercury ☿	Venus ♀	Mars ♂	Jupiter ♃	Saturn ♄	Uranus ♅	Neptune ♆	Pluto ♇
♉	♈	♊	♒	♑	♈	♓	♐
(29 R)	3 ♉			(23 R)			(R)

NB: All zodiac symbols refer to astronomical constellations, not astrological signs *(see p. 24)*

Planetary aspects
(**Bold** = *visible to naked eye)*

1
2 ☿△♇ 5^a ☾☌♇ 10^a
3 **☾☌♄** 3_p
4 **☾☌♃** 8_p
5
6 ♀△♇ 7^a ☾☌♆ 3_p
7
8
9 ♀☊ 12_p
10 ☾☌⛢ 7_p
11
12 ☿△♄ 3_p **☽●♀** 6_p
13 **☽☌☿** 3_p
14
15
16 **☽☌♂** 1^a
17 ☽☍♇ 2^a ☉△♇ 6^a
18 ☽☍♄ 10^a
19 ☽☍♃ 6_p ♀△♄ 10_p
20
21 ☽☍♆ 9^a
22
23
24 ☽☍⛢ 6_p
25
26
27 ☾☍♀ 11^a ☾☍☿ 2_p
28
29 ☿☌♀ 1^a ☾☍♂ 11^a ☾☌♇ 6_p
30 ☿☋ 12_p **☾☌♄** 11_p
31 ♂△♆ 1^a

Planet (naked eye) visibility
Evening: Mercury (to 27th), Venus, Mars
All night:
Morning: Jupiter, Saturn

May 2021

Mercury, Venus and Saturn in Earth constellations, together with two Earth trines will bring cooler weather. Uranus and Pluto bring Warmth while Mars and Jupiter bring Light influences. Only Neptune in Pisces may bring some rain.

Northern Transplanting Time
April 18 to May 1 4^a and
May 15 8_p to May 28 1_p
Southern Transplanting Time
May 1 8^a to May 15 4_p and
May 28 5_p to June 11

Transplant **table potatoes** at Root times. Transplant **seed potatoes** for 2022 from May 10 1^a to May 11 7_p.

Hay should be cut between May 14 9_p and May 17 4^a, and at other Flower times.

Control:
Moths from May 6 4_p to May 9 7_p.
Flies by burning fly papers in the cow barn at Flower times.
Mole crickets ash from May 25 8_p to May 27 8_p.
Chitinous insects, wheat weevil, Colorado beetle and varroa from May 11 8_p to May 14 8_p.

Begin **queen bee** rearing (grafting or larval transfer, comb insertion, cell punching) between May 14 9_p and May 17 4^a and at other Flower times.

Unfavorable time

June 2021

Date	Const. of Moon	Solar & lunar aspects	Trines	Moon El'ment	Parts of the plant enhanced by Moon or planets

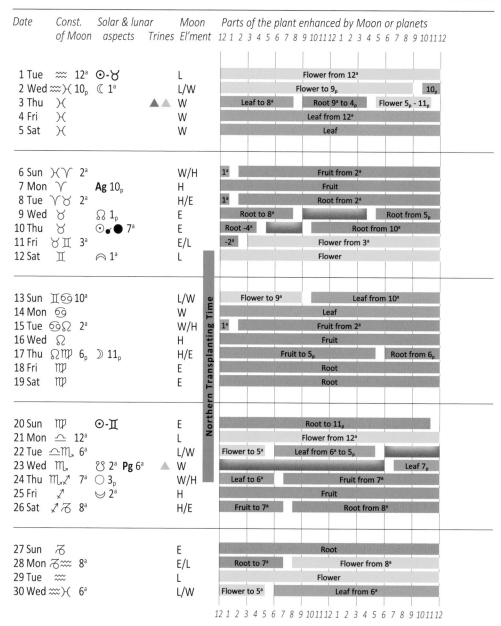

1 Tue	♒ 12ᵃ	☉-♉		L	Flower from 12ᵃ
2 Wed	♒⤓♓ 10ₚ	☾ 1ᵃ		L/W	Flower to 9ₚ / 10ₚ
3 Thu	♓		▲▲	W	Leaf to 8ᵃ / Root 9ᵃ to 4ₚ / Flower 5ₚ - 11ₚ
4 Fri	♓			W	Leaf from 12ᵃ
5 Sat	♓			W	Leaf
6 Sun	♓♈ 2ᵃ			W/H	1ᵃ Fruit from 2ᵃ
7 Mon	♈	**Ag** 10ₚ		H	Fruit
8 Tue	♈♉ 2ᵃ			H/E	1ᵃ Root from 2ᵃ
9 Wed	♉	♌ 1ₚ		E	Root to 8ᵃ / Root from 5ₚ
10 Thu	♉	☉⚹● 7ᵃ		E	Root -4ᵃ / Root from 10ᵃ
11 Fri	♉♊ 3ᵃ			E/L	-2ᵃ Flower from 3ᵃ
12 Sat	♊	⌢ 1ᵃ		L	Flower
13 Sun	♊♋ 10ᵃ			L/W	Flower to 9ᵃ / Leaf from 10ᵃ
14 Mon	♋			W	Leaf
15 Tue	♋♌ 2ᵃ			W/H	1ᵃ Fruit from 2ᵃ
16 Wed	♌			H	Fruit
17 Thu	♌♍ 6ₚ	☽ 11ₚ		H/E	Fruit to 5ₚ / Root from 6ₚ
18 Fri	♍			E	Root
19 Sat	♍			E	Root
20 Sun	♍	☉-♊		E	Root to 11ₚ
21 Mon	⚖ 12ᵃ			L	Flower from 12ᵃ
22 Tue	⚖♏ 6ᵃ			L/W	Flower to 5ᵃ / Leaf from 6ᵃ to 5ₚ
23 Wed	♏	☍ 2ᵃ **Pg** 6ᵃ	▲	W	Leaf 7ₚ
24 Thu	♏♐ 7ᵃ	◯ 3ₚ		W/H	Leaf to 6ᵃ / Fruit from 7ᵃ
25 Fri	♐	⌣ 2ᵃ		H	Fruit
26 Sat	♐♑ 8ᵃ			H/E	Fruit to 7ᵃ / Root from 8ᵃ
27 Sun	♑			E	Root
28 Mon	♑♒ 8ᵃ			E/L	Root to 7ᵃ / Flower from 8ᵃ
29 Tue	♒			L	Flower
30 Wed	♒⤓♓ 6ᵃ			L/W	Flower to 5ᵃ / Leaf from 6ᵃ

Northern Transplanting Time (spanning 12 Jun – 24 Jun)

12 1 2 3 4 5 6 7 8 9 10 11 12 1 2 3 4 5 6 7 8 9 10 11 12

Mercury ☿	Venus ♀	Mars ♂	Jupiter ♃	Saturn ♄	Uranus ♅	Neptune ♆	Pluto ♇
♉	♉ 3 ♊	♊	♒	♑	♈	♓	♐
(R 22 D)	25 ♋	7 ♋	(20 R)	(R)		(25 R)	(R)

NB: All zodiac symbols refer to astronomical constellations, not astrological signs (see p. 24)

Planetary aspects
*(**Bold** = visible to naked eye)*

1	☽☌♃ 8$_a$
2	
3	☽☌♆ 1$_a$ ☉△♄ 3$_p$ ♀△♃ 8$_p$
4	
5	♂☍♇ 4$_p$
6	
7	☽☌☊ 4$_a$
8	
9	
10	☽☌☿ 9$_a$ ☉☌☿ 9$_p$
11	
12	☽☌♀ 3$_a$
13	☽☍♇ 7$_a$ ☽☌♂ 5$_p$
14	☽☍♄ 3$_p$
15	
16	☽☍♃ 3$_a$
17	☽☍♆ 5$_p$
18	
19	
20	
21	☽☍☊ 6$_a$ ♀△♆ 10$_a$
22	
23	☉△♃ 6$_a$ ☽☍☿ 11$_a$ ♀☍♇ 8$_p$
24	
25	
26	☽☌♇ 4$_a$ ☽☍♀ 9$_a$
27	☽☍♂ 3$_a$ ☽☌♄ 7$_a$
28	☽☌♃ 6$_p$
29	
30	☽☌♆ 8$_a$

Planet (naked eye) visibility
Evening: Venus, Mars (to 29th)
All night: -
Morning: Jupiter, Saturn

June 2021

There is a mixed picture for June. There are Warmth, Light and Earth constellations. Pluto and Uranus are in Warmth constellations, while Mercury and Saturn are in Earth constellations, intensifying the cold effect in their retrograde motion. Jupiter and (for most of the month) Venus are in Light constellations, and there are two Light trines. Mars moves into Cancer on June 7, and, together with Neptune in Pisces, may bring precipitation.

Northern Transplanting Time
June 12 2$_a$ to June 24 11$_p$
Southern Transplanting Time
May 28 to June 11 10$_p$ and
June 25 4$_a$ to July 9

Cut **hay** at Flower times.

Begin **queen bee** rearing at Fruit and Flower times, avoiding unfavorable times.

Control:
Chitinous insects, wheat weevil, Colorado beetle and varroa from June 8 2$_a$ to June 11 2$_a$.
Flies by burning fly papers in the cow barn from June 1 1$_a$ to June 2 9$_p$, and June 28 8$_a$ to June 30 5$_a$ and at other Flower times.
Grasshoppers from June 11 3$_a$ to June 13 9$_a$.
Mole crickets ash from June 22 6$_a$ to June 24 6$_a$.

Maria Thun's biodynamic tree log preparations
Cut **oak** logs, fill them with ground **oak bark** and put them into the earth on June 5 between 5$_a$ and 10$_p$, and between June 30 10$_p$ and July 1 3$_p$.
Cut **birch** logs, fill with dried **yarrow** and put into the ground between June 23 7$_p$ and June 24 2$_a$.

June

■■■■ *Unfavorable time*   43

All times in EDT

Date	Const. of Moon	Solar & lunar aspects	Moon Trines	El'ment	Parts of the plant enhanced by Moon or planets

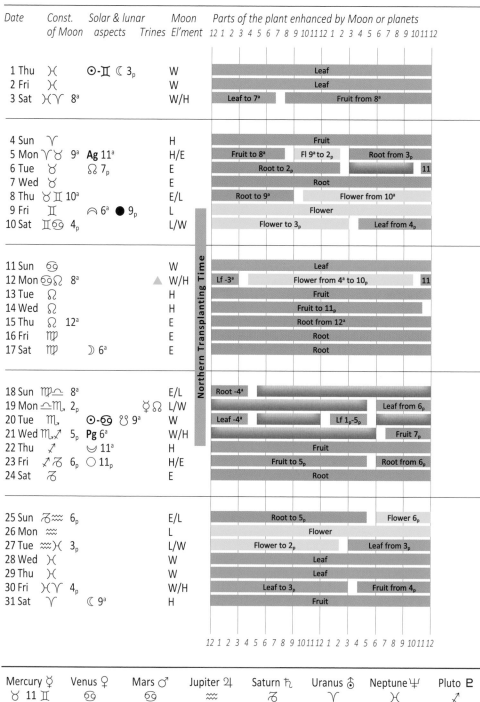

Date	Const. of Moon	Solar & lunar aspects	Moon Trines	El'ment	Parts of the plant enhanced by Moon or planets
1 Thu	♓	☉-♊ ☽ 3ₚ		W	Leaf
2 Fri	♓			W	Leaf
3 Sat	♓♈ 8ᵃ			W/H	Leaf to 7ᵃ / Fruit from 8ᵃ
4 Sun	♈			H	Fruit
5 Mon	♈♉ 9ᵃ	**Ag** 11ᵃ		H/E	Fruit to 8ᵃ / Fl 9ᵃ to 2ₚ / Root from 3ₚ
6 Tue	♉	♌ 7ₚ		E	Root to 2ₚ / 11
7 Wed	♉			E	Root
8 Thu	♉♊ 10ᵃ			E/L	Root to 9ᵃ / Flower from 10ᵃ
9 Fri	♊	⌒ 6ᵃ ● 9ₚ		L	Flower
10 Sat	♊♋ 4ₚ			L/W	Flower to 3ₚ / Leaf from 4ₚ
11 Sun	♋			W	Leaf
12 Mon	♋♌ 8ᵃ		▲	W/H	Lf -3ᵃ / Flower from 4ᵃ to 10ₚ / 11
13 Tue	♌			H	Fruit
14 Wed	♌			H	Fruit to 11ₚ
15 Thu	♌ 12ᵃ			E	Root from 12ᵃ
16 Fri	♍			E	Root
17 Sat	♍	☽ 6ᵃ		E	Root
18 Sun	♍♎ 8ᵃ			E/L	Root -4ᵃ
19 Mon	♎♏ 2ₚ	☿ ♌		L/W	Leaf from 6ₚ
20 Tue	♏	☉-♋ ☊ 9ᵃ		W	Leaf -4ᵃ / Lf 1ₚ-5ₚ
21 Wed	♏♐ 5ₚ	**Pg** 6ᵃ		W/H	Fruit 7ₚ
22 Thu	♐	⌣ 11ᵃ		H	Fruit
23 Fri	♐♑ 6ₚ	○ 11ₚ		H/E	Fruit to 5ₚ / Root from 6ₚ
24 Sat	♑			E	Root
25 Sun	♑♒ 6ₚ			E/L	Root to 5ₚ / Flower 6ₚ
26 Mon	♒			L	Flower
27 Tue	♒♓ 3ₚ			L/W	Flower to 2ₚ / Leaf from 3ₚ
28 Wed	♓			W	Leaf
29 Thu	♓			W	Leaf
30 Fri	♓♈ 4ₚ			W/H	Leaf to 3ₚ / Fruit from 4ₚ
31 Sat	♈	☽ 9ᵃ		H	Fruit

Northern Transplanting Time

12 1 2 3 4 5 6 7 8 9 10 11 12 1 2 3 4 5 6 7 8 9 10 11 12

Mercury ☿	Venus ♀	Mars ♂	Jupiter ♃	Saturn ♄	Uranus ♅	Neptune ♆	Pluto ♇
♉ 11 ♊	♋	♋	♒	♑	♈	♓	♐
26 ♋	12 ♌	11 ♌	(R)	(R)		(R)	(R)

NB: All zodiac symbols refer to astronomical constellations, not astrological signs (see p. 24)

Planetary aspects
(**Bold** = *visible to naked eye*)

1	♂☍♄ 9ᵃ
2	
3	
4	☾☌⊕ 1ₚ
5	
6	♀☍♄ 11ₚ
7	
8	☾☌☿ 1ᵃ
9	
10	☽☍♇ 12ₚ
11	☽☍♄ 7ₚ
12	**☽☌♀ 7ᵃ** ☽☌♂ 9ᵃ ☿△♃ 4ₚ
13	☽☍♃ 7ᵃ ♀☌♂ 10ᵃ
14	☽☍♆ 10ₚ
15	☉△♆ 5ᵃ
16	
17	☉☍♇ 7ₚ
18	☽☍⊕ 3ₚ
19	☿☊ 5ᵃ
20	
21	
22	♀☍♃ 9ᵃ
23	☽☍☿ 4ᵃ ☽☌♇ 1ₚ
24	☿△♆ 1ₚ **☽☌♄ 2ₚ**
25	☿☍♇ 4ₚ **☽☍♂ 7ₚ**
26	**☽☌♃ 1ᵃ** **☽☍♀ 9ᵃ**
27	**☽☌♆ 5ₚ**
28	
29	♂☍♃ 12ₚ
30	
31	**☽☌⊕ 10ₚ**

Planet (naked eye) visibility
Evening: Venus
All night: Jupiter, Saturn
Morning:

July 2021

In the first half of the month Venus and Mars, (as well as Neptune throughout) are in Water constellations bringing rain. Then they move into Leo; together with Uranus in Aries and Pluto in Sagittarius this will bring warmth.

Northern Transplanting Time
July 9 8ᵃ to July 22 9ᵃ
Southern Transplanting Time
June 25 to July 9 4ᵃ and July 22 1ₚ to Aug 5

Late hay cut at Flower times.

Summer harvest for seeds:
Flower plants: Harvest at Flower times, specially in the first half of the month.
 Fruit plants from July 12 11ₚ to July 14 11ₚ, or at other Fruit times.
 Harvest **leaf plants** at Leaf times.
 Harvest **root plants** at Root times, especially July 5 3ₚ to July 8 9ᵃ, and July 15 1ᵃ to July 18 4ᵃ.
 Always avoid unfavorable times.

Control
Flies: burn fly papers in the cow barn at Flower times.
Slugs: burn from July 10 4ₚ to July 12 7ᵃ. Spray leaf plants and the soil with horn silica early in the morning during Leaf times.
Grasshoppers from July 8 10ᵃ to July 10 3ₚ.

Maria Thun's biodynamic tree log preparations
Cut **oak** logs, fill with ground **oak bark** and put them into the ground between June 30 10ₚ and July 1 3ₚ or July 29 between 1ᵃ and 8ₚ.
 Cut **maple** logs, fill with dried **dandelion** and put them into the ground between July 17 8ᵃ and July 18 1ᵃ.
 Cut **birch** logs, fill with dried **yarrow** and put them into the ground between July 21 10ₚ and July 22 3ₚ.
 Cut **larch** logs, fill with dried **camomile** and put them into the ground on July 25 between 5ᵃ and 10ₚ.

Date	Const. of Moon	Solar & lunar aspects	Moon Trines	El'ment	Parts of the plant enhanced by Moon or planets

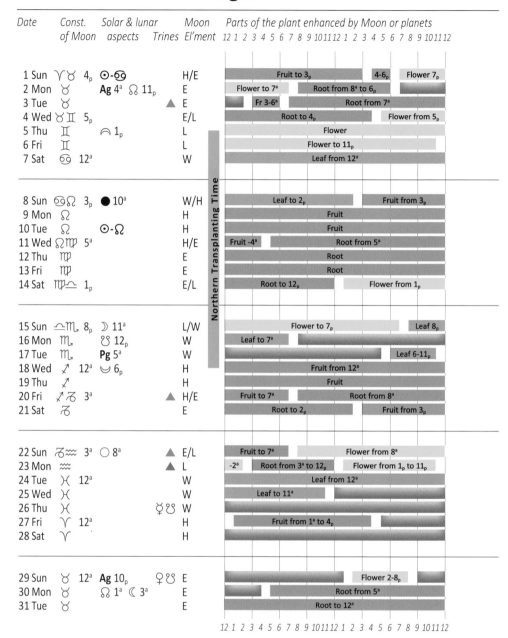

1 Sun ♈♉ 4ₚ	☉-♋		H/E	Fruit to 3ₚ / 4-6ₚ / Flower 7ₚ
2 Mon ♉	Ag 4ª ☋ 11ₚ		E	Flower to 7ª / Root from 8ª to 6ₚ
3 Tue ♉		▲	E	Fr 3-6ª / Root from 7ª
4 Wed ♉♊ 5ₚ			E/L	Root to 4ₚ / Flower from 5ₚ
5 Thu ♊	⌒ 1ₚ		L	Flower
6 Fri ♊			L	Flower to 11ₚ
7 Sat ♋ 12ª			W	Leaf from 12ª

Northern Transplanting Time

8 Sun ♋♌ 3ₚ	● 10ª		W/H	Leaf to 2ₚ / Fruit from 3ₚ
9 Mon ♌			H	Fruit
10 Tue ♌	☉-♌		H	Fruit
11 Wed ♌♍ 5ª			H/E	Fruit -4ª / Root from 5ª
12 Thu ♍			E	Root
13 Fri ♍			E	Root
14 Sat ♍♎ 1ₚ			E/L	Root to 12ₚ / Flower from 1ₚ

15 Sun ♎♏ 8ₚ	☽ 11ª		L/W	Flower to 7ₚ / Leaf 8ₚ
16 Mon ♏	☋ 12ₚ		W	Leaf to 7ª
17 Tue ♏	Pg 5ª		W	Leaf 6-11ₚ
18 Wed ♐ 12ª	⌣ 6ₚ		H	Fruit from 12ª
19 Thu ♐			H	Fruit
20 Fri ♐♑ 3ª		▲	H/E	Fruit to 7ª / Root from 8ª
21 Sat ♑			E	Root to 2ₚ / Fruit from 3ₚ

22 Sun ♑♒ 3ª	○ 8ª	▲	E/L	Fruit to 7ª / Flower from 8ª
23 Mon ♒		▲	L	-2ª / Root from 3ª to 12ₚ / Flower from 1ₚ to 11ₚ
24 Tue ♓ 12ª			W	Leaf from 12ª
25 Wed ♓			W	Leaf to 11ª
26 Thu ♓	☿☋		W	
27 Fri ♈ 12ª			H	Fruit from 1ª to 4ₚ
28 Sat ♈			H	

29 Sun ♉ 12ª	Ag 10ₚ	♀☋ E	Flower 2-8ₚ
30 Mon ♉	☋ 1ª ☾ 3ª	E	Root from 5ª
31 Tue ♉		E	Root to 12ª

12 1 2 3 4 5 6 7 8 9 10 11 12 1 2 3 4 5 6 7 8 9 10 11 12

Mercury ☿	Venus ♀	Mars ♂	Jupiter ♃	Saturn ♄	Uranus ⛢	Neptune ♆	Pluto ♇
♋ 5 ♌	♌	♌	♒ 24 ♑	♑	♈	♓	♐
25 ♍	10 ♍		(R)	(R)	(19 R)	(R)	(R)

NB: All zodiac symbols refer to astronomical constellations, not astrological signs (see p. 24)

Planetary aspects
(**Bold** = *visible to naked eye)*

1 ☉☌☿ 10^a ☿☍♄ 6_p
2 ☉☍♄ 2^a
3 ♀△⛢ 3^a
4
5
6 ☾☍♇ 6_p
7 ☾☍♄ 10_p

8
9 ☽☌☿ 2^a ☽☍♃ 8^a ♀☍♆ 8_p
10 ☽☌♂ 1^a ☿☍♃ 9_p
11 ☽☍♆ 3^a **☽☌♀** 6^a ♀△♇ 7_p
12
13
14 ☽☍⛢ 9_p

15
16
17
18 ☿☌♂ 11_p
19 ☽☌♇ 8_p ☉☍♃ 8_p
20 ☿△⛢ 4^a **☽☌♄** 8_p
21

22 **☽☌♃** 3^a ♂△⛢ 3^a
23 ♀△♄ 9^a ☾☍♂ 1_p ☾☍☿ 10_p
24 ☾☌♆ 1^a ☿☍♆ 9_p
25 ☾☍♀ 12_p
26 ☿△♇ 10^a ☿☋ 12_p
27
28 ☾☌⛢ 6^a

29 ♀☋ 1^a
30
31

Planet (naked eye) visibility
Evening: Venus
All night: Jupiter, Saturn
Morning:

August 2021

August is dominated by Warmth constellations, reinforced by three Warmth trines. By the end of the month Mercury, Venus and Jupiter join Saturn in cooler Earth constellations. Only Neptune in Pisces will hopefully bring some rain.

Northern Transplanting Time
Aug 5 3_p to Aug 18 4_p
Southern Transplanting Time
July 22 to Aug 5 11^a and Aug 18 8_p to Sep 1

Harvest **seeds of fruit plants** and **grain** to be used for seed from Aug 8 3_p to Aug 11 4^a, and at other Fruit times, avoiding unfavorable times.

Immediately after harvest, sow catch crops like lupins, phacelia, mustard or wild flax.

Seeds for leaf plants: harvest at Leaf times, specially in the second half of the month.

Seeds for flower plants: at Flower times, specially in the second half of the month.

Burn **fly papers** in the cow barn at Flower times.

Ants in the house: burn when the Moon is in Leo, Aug 8 3_p to Aug 11 4^a.

Biodynamic preparations
Cut **yarrow** in the mornings at Fruit times after Aug 11. The blossoms should show some seed formation.

Maria Thun's biodynamic tree log preparations
Cut **larch** logs, fill with dried **camomile** and put them into the ground on Aug 1 between 7^a and 11_p or between Aug 10 10^a and Aug 11 3^a.

Cut **maple** logs, fill with dried **dandelion** and put them into the ground between Aug 1 3_p and Aug 2 8^a or between Aug 19 9^a and Aug 20 2^a.

Cut **birch** logs, fill with dried **yarrow** (from last year) and put them into the ground between Aug 9 9^a and Aug 10 2^a.

Unfavorable time

September 2021

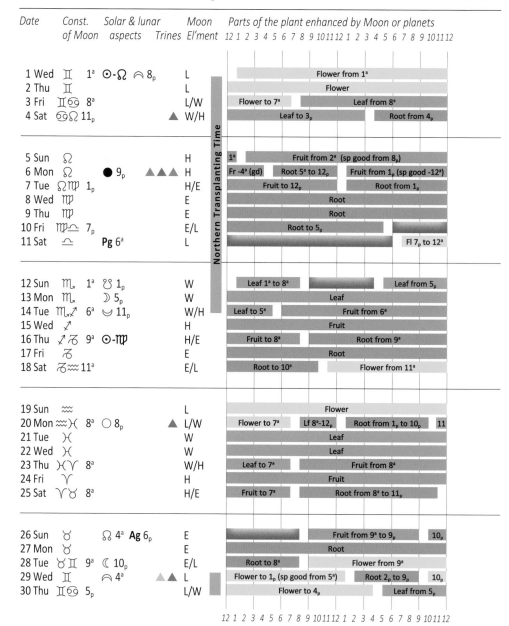

Date	Const. of Moon	Solar & lunar aspects	Trines	Moon El'ment	Parts of the plant enhanced by Moon or planets 12 1 2 3 4 5 6 7 8 9 10 11 12 1 2 3 4 5 6 7 8 9 10 11 12
1 Wed	♊	1ᵃ ☉-♌ ⌢ 8ₚ		L	Flower from 1ᵃ
2 Thu	♊			L	Flower
3 Fri	♊♋ 8ᵃ			L/W	Flower to 7ᵃ · Leaf from 8ᵃ
4 Sat	♋♌ 11ₚ		▲	W/H	Leaf to 3ₚ · Root from 4ₚ
5 Sun	♌			H	1ᵃ Fruit from 2ᵃ (sp good from 8ₚ)
6 Mon	♌	● 9ₚ	▲▲▲	H	Fr -4ᵃ (gd) · Root 5ᵃ to 12ₚ · Fruit from 1ₚ (sp good -12ᵃ)
7 Tue	♌♍ 1ₚ			H/E	Fruit to 12ₚ · Root from 1ₚ
8 Wed	♍			E	Root
9 Thu	♍			E	Root
10 Fri	♍⌐ 7ₚ			E/L	Root to 5ₚ
11 Sat	⌐	Pg 6ᵃ		L	Fl 7ₚ to 12ᵃ
12 Sun	♏	1ᵃ ♋ 1ₚ		W	Leaf 1ᵃ to 8ᵃ · Leaf from 5ₚ
13 Mon	♏	☽ 5ₚ		W	Leaf
14 Tue	♏♐ 6ᵃ	♋ 11ₚ		W/H	Leaf to 5ᵃ · Fruit from 6ᵃ
15 Wed	♐			H	Fruit
16 Thu	♐♑ 9ᵃ	☉-♍		H/E	Fruit to 8ᵃ · Root from 9ᵃ
17 Fri	♑			E	Root
18 Sat	♑♒ 11ᵃ			E/L	Root to 10ᵃ · Flower from 11ᵃ
19 Sun	♒			L	Flower
20 Mon	♒♓ 8ᵃ	○ 8ₚ	▲	L/W	Flower to 7ᵃ · Lf 8ᵃ-12ₚ · Root from 1ₚ to 10ₚ · 11
21 Tue	♓			W	Leaf
22 Wed	♓			W	Leaf
23 Thu	♓♈ 8ᵃ			W/H	Leaf to 7ᵃ · Fruit from 8ᵃ
24 Fri	♈			H	Fruit
25 Sat	♈♉ 8ᵃ			H/E	Fruit to 7ᵃ · Root from 8ᵃ to 11ₚ
26 Sun	♉	♋ 4ᵃ Ag 6ₚ		E	Fruit from 9ᵃ to 9ₚ · 10ₚ
27 Mon	♉			E	Root
28 Tue	♉♊ 9ᵃ	☾ 10ₚ		E/L	Root to 8ᵃ · Flower from 9ᵃ
29 Wed	♊	⌢ 4ᵃ	▲▲	L	Flower to 1ₚ (sp good from 5ᵃ) · Root 2ₚ to 9ₚ · 10ₚ
30 Thu	♊♋ 5ₚ			L/W	Flower to 4ₚ · Leaf from 5ₚ

12 1 2 3 4 5 6 7 8 9 10 11 12 1 2 3 4 5 6 7 8 9 10 11 12

Northern Transplanting Time

Sep

Mercury ☿	Venus ♀	Mars ♂	Jupiter ♃	Saturn ♄	Uranus ⛢	Neptune ♆	Pluto ♇
♍	♍	♌	♑	♑	♈	♓ 18 ♒	♐
(27 R)	19 ⌐		(R)	(R)	(R)	(R)	(R)

NB: All zodiac symbols refer to astronomical constellations, not astrological signs (see p. 24)

September 2021

1	
2	♂☌♆ 2ₚ
3	☽☍♇ 2ᵃ
4	☽☍♄ 3ᵃ ☿△♄ 10ₚ
5	☽☍♃ 10ᵃ
6	♂△♇ 8ᵃ ♀△♃ 9ᵃ ☉△☋ 9ₚ
7	☽☍♆ 10ᵃ ☽☌♂ 3ₚ
8	☽☌☿ 9ₚ
9	
10	☽☌♀ 1ᵃ
11	☽☍☋ 3ᵃ
12	
13	
14	☉☍♆ 5ᵃ
15	
16	☽☌♇ 2ᵃ ☉△♇ 10ₚ
17	☽☌♄ 1ᵃ
18	☽☌♃ 5ᵃ
19	
20	☽☌♆ 8ᵃ ☿△♃ 7ₚ
21	☽☍♂ 7ᵃ
22	☽☍☿ 10ₚ
23	♀☍☋ 6ᵃ
24	☽☌☋ 1ₚ ☽☍♀ 4ₚ
25	♂△♄ 6ₚ
26	
27	
28	
29	♀△♆ 12ₚ ☉△♄ 6ₚ
30	☽☍♇ 10ᵃ

Planet (naked eye) visibility
Evening: Venus
All night: Jupiter, Saturn
Morning:

In September there is tension between Earth constellations (with Jupiter and Saturn, and, in the first part of the month, Mercury and Venus), and Warmth constellations with Mars, Uranus and Pluto. There are two Warmth trines and four Earth trines during the month. Only Neptune for the first half of the month is in Watery Pisces.

Northern Transplanting Time
Sep 1 10ₚ to Sep 14 10ₚ and
Sep 29 6ᵃ to Oct 12
Southern Transplanting Time
Aug 19 to Sep 1 6ₚ and Sep 15 2ᵃ to Sep 29 2ᵃ

The times recommended for the **fruit harvest** are those in which the Moon is in Aries or Sagittarius (Sep 14 6ᵃ to Sep 16 8ᵃ, Sep 23 8ᵃ to Sep 25 7ᵃ) or other Fruit times.

The harvest of **root crops** is always best undertaken at Root times. Storage trials of onions, carrots, beetroot and potatoes have demonstrated this time and again.

Good times for **sowing winter grain** are when the Moon is in Leo or Sagittarius (Sep 4 11ₚ to Sep 6 4ᵃ, and Sep 6 1ₚ to Sep 7 12ₚ, and Sep 14 6ᵃ to Sep 16 8ᵃ) avoiding unfavorable times, and at other Fruit times.

Rye can if necessary also be sown at Root times with all subsequent cultivations being carried out at Fruit times.

Control slugs by burning between Sep 3 8ᵃ and Sep 4 10ₚ or between Sep 30 5ₚ and Oct 2 7ᵃ.

Maria Thun's biodynamic tree log preparations
Cut **oak** logs, fill with ground **oak bark** and put them into the ground on Sep 2 between 3ᵃ and 8ₚ.

Cut **maple** logs, fill with dried **dandelion** and put them into the ground between Sep 13 6ₚ and Sep 14 11ᵃ.

Cut **birch** logs, fill with dried **yarrow** and put them into the ground between Sep 22 7ₚ and Sep 23 12ₚ.

Sep

October 2021

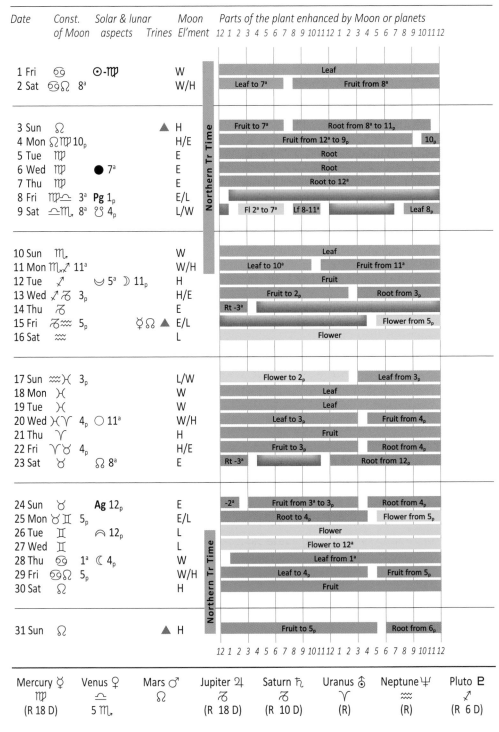

Date	Const. of Moon	Solar & lunar aspects	Moon Trines	Moon El'ment	Parts of the plant enhanced by Moon or planets

Parts of the plant enhanced by Moon or planets: 12 1 2 3 4 5 6 7 8 9 10 11 12 1 2 3 4 5 6 7 8 9 10 11 12

1 Fri ♋ ☉-♍ — W — Leaf
2 Sat ♋♌ 8ᵃ — W/H — Leaf to 7ᵃ / Fruit from 8ᵃ

3 Sun ♌ ▲ H — Fruit to 7ᵃ / Root from 8ᵃ to 11ₚ
4 Mon ♌♍ 10ₚ — H/E — Fruit from 12ᵃ to 9ₚ / 10ₚ
5 Tue ♍ — E — Root
6 Wed ♍ ● 7ᵃ — E — Root
7 Thu ♍ — E — Root to 12ᵃ
8 Fri ♍⟋ 3ᵃ **Pg** 1ₚ — E/L — Fl 2ᵃ to 7ᵃ / Lf 8-11ᵃ / Leaf 8ₚ
9 Sat ⟋♏ 8ᵃ ☋ 4ₚ — L/W

Northern Tr Time

10 Sun ♏ — W — Leaf
11 Mon ♏♐ 11ᵃ — W/H — Leaf to 10ᵃ / Fruit from 11ᵃ
12 Tue ♐ ☌ 5ᵃ ☽ 11ₚ — H — Fruit
13 Wed ♐♑ 3ₚ — H/E — Fruit to 2ₚ / Root from 3ₚ
14 Thu ♑ — E — Rt -3ᵃ
15 Fri ♑♒ 5ₚ ☿☋ ▲ E/L — Flower from 5ₚ
16 Sat ♒ — L — Flower

17 Sun ♒♓ 3ₚ — L/W — Flower to 2ₚ / Leaf from 3ₚ
18 Mon ♓ — W — Leaf
19 Tue ♓ — W — Leaf
20 Wed ♓♈ 4ₚ ○ 11ᵃ — W/H — Leaf to 3ₚ / Fruit from 4ₚ
21 Thu ♈ — H — Fruit
22 Fri ♈♉ 4ₚ — H/E — Fruit to 3ₚ / Root from 4ₚ
23 Sat ♉ ☋ 8ᵃ — E — Rt -3ᵃ / Root from 12ₚ

24 Sun ♉ **Ag** 12ₚ — E — -2ᵃ / Fruit from 3ᵃ to 3ₚ / Root from 4ₚ
25 Mon ♉♊ 5ₚ — E/L — Root to 4ₚ / Flower from 5ₚ
26 Tue ♊ ⌢ 12ₚ — L — Flower
27 Wed ♊ — L — Flower to 12ᵃ
28 Thu ♋ 1ᵃ ☾ 4ₚ — W — Leaf from 1ᵃ
29 Fri ♋♌ 5ₚ — W/H — Leaf to 4ₚ / Fruit from 5ₚ
30 Sat ♌ — H — Fruit

Northern Tr Time

31 Sun ♌ ▲ H — Fruit to 5ₚ / Root from 6ₚ

12 1 2 3 4 5 6 7 8 9 10 11 12 1 2 3 4 5 6 7 8 9 10 11 12

Mercury ☿	Venus ♀	Mars ♂	Jupiter ♃	Saturn ♄	Uranus ♅	Neptune ♆	Pluto ♇
♍	⌢	♌	♑	♑	♈	♒	♐
(R 18 D)	5♏		(R 18 D)	(R 10 D)	(R)	(R)	(R 6 D)

NB: All zodiac symbols refer to astronomical constellations, not astrological signs (see p. 24)

Planetary aspects
(**Bold** = *visible to naked eye*)

1	☾☍♄ 10ᵃ
2	☾☍♃ 3ₚ
3	☿△♃ 8ₚ
4	☾☍♆ 6ₚ
5	
6	☽♂♂ 8ᵃ ☽♂☿ 6ₚ
7	
8	☉♂♂ 1ᵃ ☽☍♁ 9ᵃ
9	☉♂☿ 12ₚ ☽♂♀ 4ₚ ☿♂♂ 7ₚ
10	
11	
12	
13	☽♂♇ 7ᵃ
14	**☽♂♄ 5ᵃ**
15	☿☍ 4ᵃ ☉△♃ 8ᵃ **☽♂♃ 9ᵃ**
16	
17	☽♂♆ 1ₚ
18	♂△♃ 11ₚ
19	☽☍☿ 1ᵃ
20	☽☍♂ 2ᵃ
21	☾♂♁ 6ₚ
22	
23	
24	☾☍♀ 6ₚ
25	
26	
27	☾☍♇ 6ₚ
28	☾☍♄ 7ₚ
29	
30	☾☍♃ 1ᵃ
31	☿△♃ 11ₚ

October continues in the tension between Mercury, Jupiter and Saturn in Earth constellations, reinforced by three Earth trines, and Mars, Uranus and Pluto in Warmth constellations. Neptune is in Light Aquarius and Venus, after Oct 5, is in Watery Scorpio.

Northern Transplanting Time
Sep 29 to Oct 12 3ᵃ and
Oct 26 2ₚ to Nov 8
Southern Transplanting Time
Oct 12 7ᵃ to Oct 26 10ᵃ

Store fruit at any Fruit or Flower time outside transplanting time.

Harvest seeds of root plants at Root times, **seeds for leaf plants** at Leaf times, and **seeds for flower plants** at Flower times.

All **cleared ground** should be treated with compost and sprayed with barrel preparation, and plowed ready for winter.

Control slugs by burning between Sep 30 5ₚ and Oct 2 7ᵃ.

Burn feathers or skins of **warm blooded pests** from Oct 22 4ₚ to Oct 25 4ₚ. Ensure the fire is glowing hot (don't use grilling charcoal). Lay dry feathers or skins on the glowing embers. After they have cooled, collect the light grey ash and grind for an hour with a pestle and mortar. This increases their efficacy and the ash can be potentized later. *The burning and grinding should be completed by Oct 25 4ₚ.*

Planet (naked eye) visibility
Evening: Venus, Saturn
All night: Jupiter
Morning: Mercury (from 16th)

Oct

▬▬▬ *Unfavorable time* 51

November 2021

Date	Const. of Moon	Solar & lunar aspects	Trines	Moon El'ment	Parts of the plant enhanced by Moon or planets

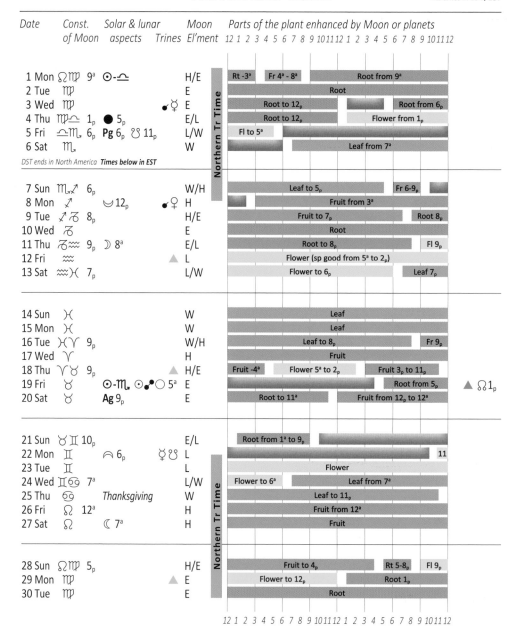

Date	Const. of Moon	Solar & lunar aspects	Trines	Moon El'ment
1 Mon	♌︎♍︎ 9ᵃ	☉-♎︎		H/E
2 Tue	♍︎			E
3 Wed	♍︎		⚹☿	E
4 Thu	♍︎♎︎ 1ₚ	● 5ₚ		E/L
5 Fri	♎︎♏︎ 6ₚ	**Pg** 6ₚ ☋ 11ₚ		L/W
6 Sat	♏︎			W

DST ends in North America **Times below in EST**

7 Sun	♏︎♐︎ 6ₚ			W/H
8 Mon	♐︎	☌ 12ₚ	⚹♀	H
9 Tue	♐︎♑︎ 8ₚ			H/E
10 Wed	♑︎			E
11 Thu	♑︎♒︎ 9ₚ	☽ 8ᵃ		E/L
12 Fri	♒︎		▲	L
13 Sat	♒︎♓︎ 7ₚ			L/W

14 Sun	♓︎			W
15 Mon	♓︎			W
16 Tue	♓︎♈︎ 9ₚ			W/H
17 Wed	♈︎			H
18 Thu	♈︎♉︎ 9ₚ		▲	H/E
19 Fri	♉︎	☉-♏︎, ☉●☉○ 5ᵃ		E
20 Sat	♉︎	**Ag** 9ₚ		E

21 Sun	♉︎♊︎ 10ₚ			E/L
22 Mon	♊︎	⌢ 6ₚ	☿☋	L
23 Tue	♊︎			L
24 Wed	♊︎♋︎ 7ᵃ			L/W
25 Thu	♋︎	*Thanksgiving*		W
26 Fri	♌︎ 12ᵃ			H
27 Sat	♌︎	☾ 7ᵃ		H

28 Sun	♌︎♍︎ 5ₚ			H/E
29 Mon	♍︎		▲	E
30 Tue	♍︎			E

Mercury ☿	Venus ♀	Mars ♂	Jupiter ♃	Saturn ♄	Uranus ♅	Neptune ♆	Pluto ♇
♍︎ 11♎︎	♏︎	♌︎	♑︎	♑︎	♈︎	♒︎	♐︎
22♏︎	3 ♐︎	13 ♎︎			(R)	(R)	

NB: All zodiac symbols refer to astronomical constellations, not astrological signs (see p. 24)

Planetary aspects
(**Bold** = *visible to naked eye*)

1	☾☍♆ 3ᵃ
2	
3	**☾•☿ 3ₚ**
4	☾♂♂ 2ᵃ ☽☍⊕ 5ₚ ☉☍⊕ 8ₚ
5	
6	
7	
8	☽•♀ 1ᵃ
9	☽♂♇ 1ₚ
10	☿♂♂ 8ᵃ ☽♂♄ 11ᵃ
11	☽♂♃ 3ₚ
12	☉△♆ 11ᵃ
13	☿☍⊕ 11ᵃ ☽♂♆ 5ₚ
14	
15	
16	
17	♂☍⊕ 12ₚ ☽♂⊕ 10ₚ ☽☍♂ 10ₚ
18	☿△♆ 11ᵃ ☽☍☿ 3ₚ
19	♀△⊕ 1ᵃ
20	
21	
22	☿♋ 10ᵃ
23	☾☍♀ 6ᵃ
24	☾☍♇ 1ᵃ
25	☾☍♄ 4ᵃ
26	☾☍♃ 11ᵃ
27	
28	☾☍♆ 11ᵃ
29	☉♂☿ 1ᵃ ♂△♆ 9ᵃ
30	

Jupiter and Saturn (as well as Mercury until Nov 11) are in Earth constellations, balanced by Uranus, Pluto and Venus (after Nov 3) in Warmth constellations. There are two Light trines and for a time Mercury and Mars are in Light constellations, which may ensure some brightness.

Northern Transplanting Time
Oct 26 to Nov 8 10ᵃ and
Nov 22 8ₚ to Dec 5
Southern Transplanting Time
Nov 8 2ₚ to Nov 22 4ₚ

The Flower times in Transplanting Time are ideal for **planting flower bulbs,** showing vigorous growth and vivid colors. The remaining Flower times should only be considered as back up, as bulbs planted on those times will not flower so freely.

If not already completed in October, all organic waste materials should be gathered and made into a **compost.** Applying the biodynamic preparations to the compost will ensure a rapid transformation and good fungal development. An application of barrel preparation will also help the composting process.

Fruit and forest trees will also benefit at this time from a spraying of horn manure and/ or barrel preparation when being transplanted.

Best times for **cutting Advent greenery** and **Christmas trees** for transporting are Flower times, avoiding unfavorable times.

Burn **fly papers** in cow barn at Flower times.

Planet (naked eye) visibility
Evening: Venus, Jupiter, Saturn
All night:
Morning: Mercury (to 12th), Mars (from 26th)

Nov

▮▮▮ *Unfavorable time*

December 2021

Date	Const. of Moon	Solar & lunar aspects	Moon Trines	El'ment	Parts of the plant enhanced by Moon or planets 12 1 2 3 4 5 6 7 8 9 10 11 12 1 2 3 4 5 6 7 8 9 10 11 12
1 Wed	♍︎♎︎ 11ₚ	☉-♏︎		E/L	
2 Thu	♎︎	♂		L	
3 Fri	♎︎♏︎ 4ª	☍ 10ª		L/W	
4 Sat	♏︎	☉●3ª Pg5ª ☿		W	
5 Sun	♏︎♐︎ 5ª	☋ 10ₚ		W/H	
6 Mon	♐︎			H	
7 Tue	♐︎♑︎ 5ª			H/E	
8 Wed	♑︎			E	
9 Thu	♑︎♒︎ 4ª			E/L	
10 Fri	♒︎	☽ 9ₚ		L	
11 Sat	♓︎ 1ª			W	
12 Sun	♓︎			W	
13 Mon	♓︎			W	
14 Tue	♓︎♈︎ 3ª			W/H	
15 Wed	♈︎			H	
16 Thu	♈︎♉︎ 3ª	☌ 7ₚ		H/E	
17 Fri	♉︎	Ag 9ₚ		E	
18 Sat	♉︎	○ 11ₚ		E	
19 Sun	♉︎♊︎ 4ª 11ₚ	♂☋		E/L	
20 Mon	♊︎	☉-♐︎ ♀☋	▲	L	
21 Tue	♊︎♋︎ 1ₚ			L/W	
22 Wed	♋︎			W	
23 Thu	♋︎♌︎ 6ª			W/H	
24 Fri	♌︎			H	
25 Sat	♌︎	Christmas		H	
26 Sun	♍︎ 12ª	☾ 9ₚ		E	
27 Mon	♍︎			E	
28 Tue	♍︎			E	
29 Wed	♍︎♎︎ 9ª			E/L	
30 Thu	♎︎♏︎ 3ₚ	☍ 8ₚ		L/W	
31 Fri	♏︎			W	

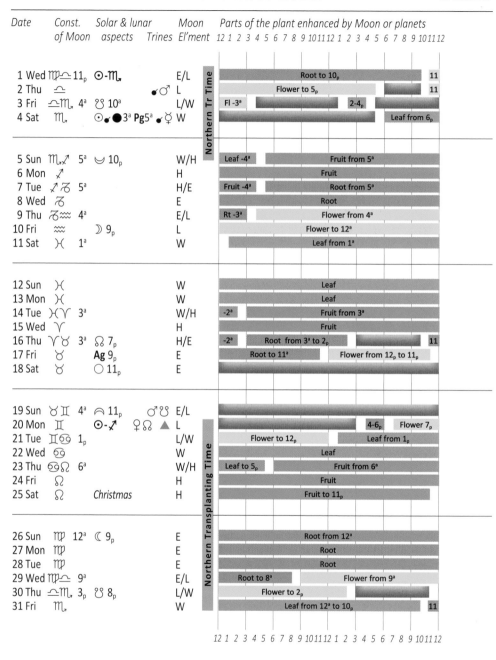

Northern Tr Time

Northern Transplanting Time

12 1 2 3 4 5 6 7 8 9 10 11 12 1 2 3 4 5 6 7 8 9 10 11 12

Mercury ☿	Venus ♀	Mars ♂	Jupiter ♃	Saturn ♄	Uranus ♅	Neptune ♆	Pluto ♇
♏︎	♐︎	♎︎	♑︎	♑︎	♈︎	♒︎	♐︎
12 ♐︎	(19 R)	9 ♏︎	9 ♒︎		(R)	(R 1 D)	

NB: All zodiac symbols refer to astronomical constellations, not astrological signs (see p. 24)

Dec

Planetary aspects
(**Bold** = *visible to naked eye)*

1
2 ☾☍♅ 2^a **☾●♂ 8_p**
3
4 ☽●☿ 8^a
5
6 **☽☌♀ 8_p** ☽☌♇ 11_p
7 **☽☌♄ 11_p**
8
9 **☽☌♃ 4^a**
10 ☽☌♆ 11_p
11 ♀☌♇ 11^a
12
13
14
15 ☽☌♅ 2^a
16 ☽☍♂ 9_p
17
18
19 ♂☋ 12_p
20 ☾☍☿ 1^a ♀☊ 3^a ☿△♅ 3_p
21 ☾☍♇ 8^a ☾☍♀ 10^a
22 ☾☍♄ 2_p
23
24 ☾☍♃ 2^a
25 ♀☌♇ 7^a ☾☍♆ 6_p
26
27
28
29 ☿☌♀ 5^a ☾☍♅ 11^a
30 ☿☌♇ 5^a
31 **☾☌♂ 3_p**

Planet (naked eye) visibility
Evening: Venus, Jupiter, Saturn
All night:
Morning: Mars

December 2021

December has a mixed picture: Jupiter and Saturn remain in the cold Earth constellation of Capricorn, but Pluto and Uranus together with Venus in the first half, and Mercury in the second half of the month are in Warmth constellations. Until Dec 9 Mars is in a Light constellation, as is Neptune for the whole month. Mercury begins in Watery Scorpio, and leaves just after Mars reaches Scorpio.

Northern Transplanting Time
Nov 22 to Dec 5 8_p and
Dec 20 2^a to Jan 2
Southern Transplanting Time
Dec 6 1^a to Dec 19 10_p

The transplanting time is good for **pruning trees and hedges.** Fruit trees should be pruned at Fruit or Flower times.

Best times for cutting **Advent greenery** and **Christmas trees** are at Flower times to ensure lasting fragrance.

We wish all our readers a blessed festive time and the best of health for the New Year of 2022

Unfavorable time

Sowing times for trees and shrubs

Sowing times depend on planetary aspects and are not specific to either northern or southern hemispheres. For trees and shrubs not mentioned here, sow at an appropriate time of the Moon's position in the zodiac, depending on the part of the tree or shrub to be enhanced. Avoid unfavorable times.

Sowing times are different from transplanting times. Seedlings should be transplanted during the descending Moon when the Moon is in a constellation corresponding to the part of the tree to to be enhanced. It is important to remember that seedlings need to sufficiently mature to withstand the winter. The time of sowing should therefore chime with local conditions and take account of the germination habit of each tree species.

Note that some species (marked in **bold**) appear in two groups.

Alder, **Apricot**, Elm, **Larch**, **Peach**:
  July 25 5ᵃ to 10ₚ;
  Aug 1 7ᵃ to 11ₚ;
  Aug 10 10ᵃ to Aug 11 3ᵃ;
  Aug 24 10ᵃ to Aug 25 3ᵃ *(also Magnolia)*;
  Nov 13 1ᵃ to 5ₚ.
Apple, **Apricot**, Copper beech, Damson, Maple, Olive, **Peach**, **Sweet chestnut**, Walnut:
  July 21 10ₚ to July 22 3ₚ;
  July 29 1ᵃ to 6ₚ;
  Aug 10 10ᵃ to Aug 11 3ᵃ;
  Aug 19 9ᵃ to Aug 20 2ᵃ.

Ash, **Cedar**, **Fir**, Hazel, **Mirabelle plum**, Rowan, **Spruce** (* = *also Hawthorn*):
  July 17 8ᵃ to July 18 1ᵃ;
  * Aug 1 3ₚ to Aug 2 8ᵃ;
  Aug 19 9ᵃ to Aug 20 2ᵃ;
  Sep 13 6ₚ to Sep 14 11ᵃ;
  * Nov 4 9ᵃ to Nov 5 2ᵃ.
Beech, **Cedar**, **Fir**, Hornbeam, Juniper, Palm, Pine, Plum, Quince, Sloe, **Spruce**, Thuja:
  June 30 10ₚ to July 1 3ₚ;
  July 6 12ₚ to July 7 5ᵃ;
  Aug 1 7ᵃ to Aug 2 8ᵃ.
Birch, **Larch**, Lime tree, **Mirabelle plum**, Pear, Robinia, Willow:
  (June 23 7ₚ to June 24 2ᵃ);
  July 6 12ₚ to July 7 5ᵃ;
  July 21 10ₚ to July 22 3ₚ;
  Aug 9 9ᵃ to Aug 10 2ᵃ *(also Magnolia)*;
  Sep 22 7ₚ to Sep 23 12ₚ.
Blackcurrant:
  Sep 22 7ₚ to Sep 23 12ₚ;
  Nov 4 9ᵃ to Nov 5 2ᵃ;
  Nov 13 1ᵃ to 5ₚ
  Nov 17 1ᵃ to 6ₚ.
Cherry, Chestnut, Horse chestnut (Buckeye), Oak, **Sweet chestnut**, Yew (* = *also Mulberry*):
  * June 5 5ᵃ to 10ₚ;
  June 30 10ₚ to July 1 3ₚ;
  July 29 1ᵃ to 6ₚ *(also Rose hip)*;
  * Sep 2 3ᵃ to 8ₚ;
  * Nov 17 1ᵃ to 6ₚ.
Lilac, Poplar, Sallow, Snowberry:
  Sep 22 7ₚ to Sep 23 12ₚ;
  Nov 13 1ᵃ to 5ₚ.

Felling times for timber

The times given are to ensure the best quality and durability of the timber.

If a large number of these trees need to be felled in the short time available, cut the bark all around the trunk to stop sap flow. The actual felling can be done later.

Those trees which are not listed should be felled at the end of the growing season at Flower times. Avoid unfavorable times.

*Alder, **Apricot**, **Elm**, **Larch**, **Peach***:
May 1 11_p to May 2 8^a;
May 12 9^a to 3_p;
July 12 4^a to 10_p;
July 24 7^a to 4_p *(also Magnolia)*;
Aug 19 5_p to Aug 20 7^a;
Sep 4 4_p to Sep 5 1^a;
Sep 20 7^a to Sep 21 1^a;
Oct 3 8^a to Oct 4 2^a;
Oct 31 12_p to Nov 1 6^a;
Nov 18 5^a to 2_p *(also Magnolia)*.

*Apple, **Apricot**, Copper beech, Damson, Maple, Olive, **Peach**, **Sweet chestnut**, Walnut*:
April 16 1_p to April 17 7^a;
June 3 8^a to June 4 2^a;
June 22 6_p to June 23 12_p;
July 12 4^a to 10_p;
Sep 5 9_p to Sep 6 3_p;
Sep 20 7^a to Sep 21 1^a;
Oct 3 8^a to Oct 4 2^a;
Oct 18 11^a to Oct 19 5^a;
Oct 31 12_p to Nov 1 6^a.

Poplar, Sallow:
Jan 13 12_p to 10_p;
Aug 19 5_p to Aug 20 7^a;
(Nov 18 6_p to 11_p).

*Ash, **Cedar**, **Fir**, Hazel, **Mirabelle plum**, Rowan, **Spruce***:
May 17 1^a to 9^a;
June 3 9^a to 6_p;
June 22 6_p to June 23 12_p;
July 14 11_p to July 15 8^a;
Sep 6 3_p to 11_p;
Sep 16 4_p to Sep 17 2^a;
Sep 29 12_p to 9_p;
Nov 12 5^a to 2_p.

*Beech, **Cedar**, **Fir**, Hornbeam, Juniper, Palm, Pine, Plum, Quince, **Spruce**, Thuja*:
March 21 11^a to March 22 2^a;
May 12 9^a to 3_p;
May 19 3_p to May 20 1^a;
June 3 9^a to 6_p;
Aug 23 2^a to 12_p;
Sep 4 4_p to Sep 5 1^a;
Sep 25 6^a to 9_p;
Sep 29 12_p to 9_p.

*Birch, **Larch**, Lime tree, **Mirabelle plum**, Pear, Robinia, Willow*:
Jan 13 12_p to 10_p;
May 6 1^a to 10^a;
May 19 3_p to May 20 1^a;
June 3 8^a to June 4 2^a;
June 21 3^a to 1_p *(also Magnolia)*;
Aug 11 12_p to 10_p;
Aug 23 2^a to 12_p;
Sep 5 9_p to Sep 6 3_p;
Sep 29 5^a to 3_p *(also Magnolia)*;
(Nov 18 6_p to 11_p.)

*Cherry, Chestnut, Horse chestnut (Buckeye), Oak, **Sweet chestnut**, Yew*:
Feb 24 9^a to 11_p;
March 21 11^a to March 22 2^a;
April 16 1_p to April 17 7^a;
Aug 21 3_p to Aug 22 6^a;
Sep 5 8_p to Sep 6 11^a;
Sep 25 6^a to 9_p;
Oct 18 11^a to Oct 19 5^a;
Nov 28 9_p to Nov 29 12_p.

The care of bees

A colony of bees lives in its hive closed off from the outside world. For extra protection against harmful influences, the inside of the hive is sealed with propolis. The link with the wider surroundings is made by the bees that fly in and out of the hive.

To make good use of cosmic rhythms, the beekeeper needs to create the right conditions in much the same way as the gardener or farmer does with the plants. The gardener works the soil and in so doing allows cosmic forces to penetrate it via the air. These forces can then be taken up and used by the plants until the soil is next moved.

When the beekeeper opens up the hive, the sealing layer of propolis is broken. This creates a disturbance, as a result of which cosmic forces can enter and influence the life of the hive until the next intervention by the beekeeper. By this means the beekeeper can directly mediate cosmic forces to the bees.

It is not insignificant which forces of the universe are brought into play when the the hive is opened. The beekeeper can consciously intervene by choosing days for working with the hive that will help the colony to develop and build up its food reserves. The bees will then reward the beekeeper by providing a portion of their harvest in the form of honey.

Earth-Root times can be selected for opening the hive if the bees need to do more building. *Light-Flower* times encourage brood activity and colony development. *Warmth-Fruit* times stimulate the collection of nectar. *Water-Leaf* times are unsuitable for working in the hive or for the removal and processing of honey.

Since the late 1970s the varroa mite has affected virtually every bee colony in Europe. Following a number of comparative trials we recommend burning and making an ash of the varroa mite in the usual way. After dynamizing it for one hour, the ash should be put in a salt-cellar and sprinkled lightly between the combs. The ash should be made and sprinkled when the Sun and Moon are in Taurus (May/June).

Over the years I have come to sprinkle small amounts of ash on the brood to strengthen its effect whenever I carry out an inspection.

Spraying the varroa ash using a salt-celler.

Biodynamic preparation plants

The biodynamic preparation plants are picked, dried, and inserted into animal sheaths (skull, bladder, etc.). *Maria Thun's log tree tree preparations* are made by inserting the plants into logs of certain trees instead of animal sheaths. The time of cutting the logs and inserting the dried plants is more critical than for traditional preparations, and therefore are given in the calendar. Both kinds are buried over winter to make the biodynamic compost preparations.

Pick *dandelions* in the morning at Flower times as soon as they are open and while the centre of the flowers are still tightly packed.

Pick *yarrow* at Fruit times when the Sun is in Leo (around the middle of August).

Pick *camomile* at Flower times just before midsummer. If they are harvested too late, seeds will begin to form and there are often grubs in the hollow heads.

Collect *stinging nettles* when the first flowers are opening, usually around midsummer. Harvest the whole plants without roots at Flower times.

Pick *valerian* at Flower times around midsummer.

All the flowers (except valerian) should be laid out on paper and dried in the shade.

Collect *oak bark* at Root times. The pithy material below the bark should not be used.

Fungal problems

The function of fungus in nature is to break down dying organic materials. It appears amongst our crops when unripe manure compost or uncomposted animal by-products such as horn and bone meal are used but also when seeds are harvested during unfavorable constellations: according to Steiner, 'When Moon forces are working too strongly on the Earth ...'

Tea can be made from horsetail *(Equisetum arvense)* and sprayed on to the soil where affected plants are growing. This draws the fungal level back down into the ground where it belongs.

The plants can be strengthened by spraying stinging nettle tea on the leaves. This will promote good assimilation, stimulate the flow of sap and help fungal diseases to disappear.

Moon diagrams

The diagrams overleaf show for each month the daily position (evenings GMT) of the Moon against the stars and other planets. For viewing in the southern hemisphere, turn the diagrams upside down.

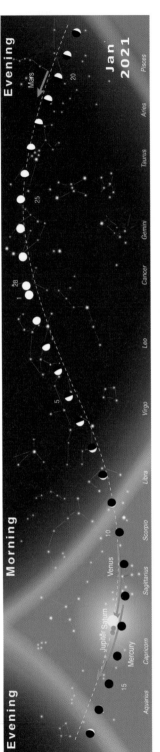

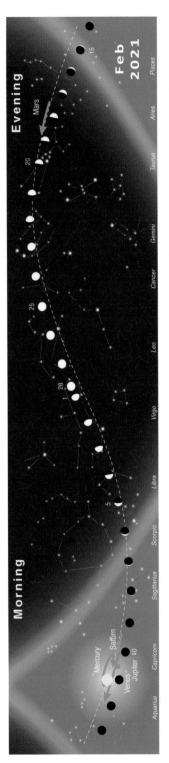

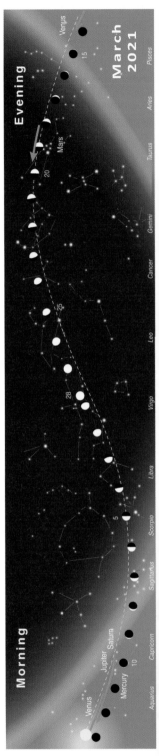

60

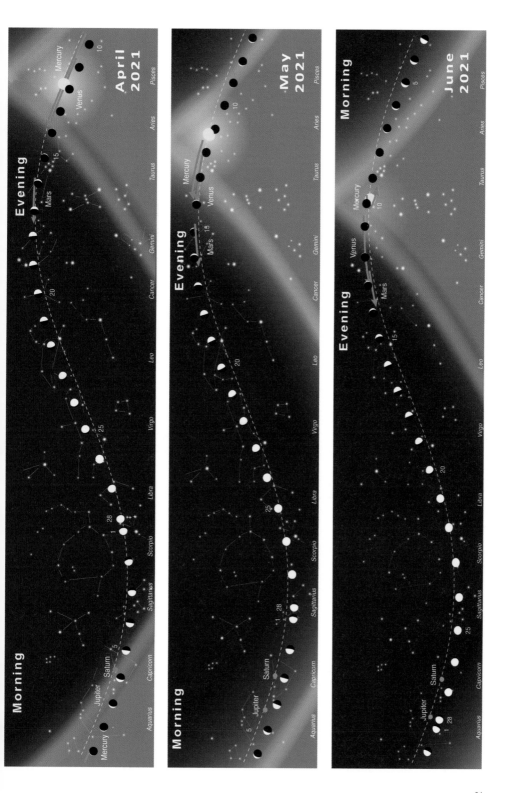

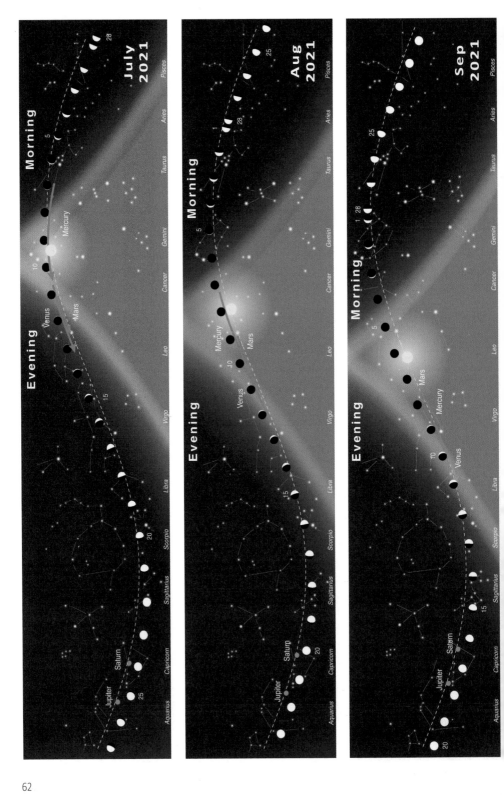

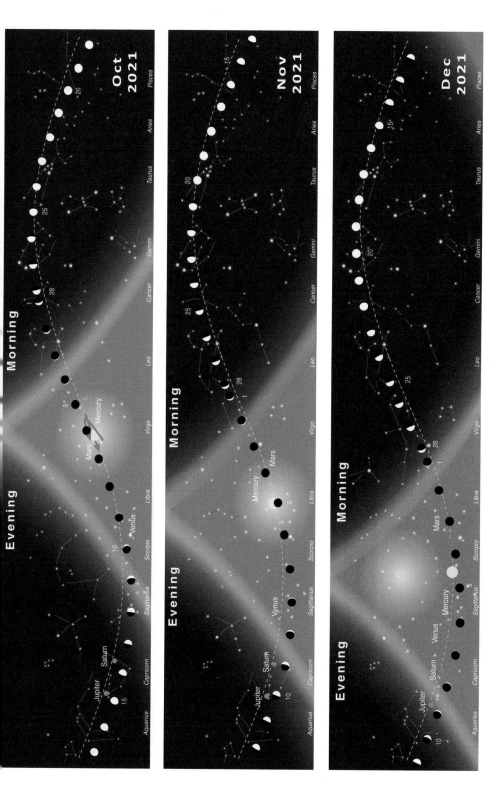

Further Reading

Berrevoets, Erik, *Wisdom of Bees: Principles of Biodynamic Beekeeping,* SteinerBooks, USA

Colquhoun, Margaret and Axel Ewald, *New Eyes for Plants,* Hawthorn

Karlsson, Britt and Per, *Biodynamic, Organic and Natural Winemaking,* Floris

Keyserlink, Adalbert Count von, *The Birth of a New Agriculture,* Temple Lodge

—, *Developing Biodynamic Agriculture,* Temple Lodge

Klett, Manfred, *Principles of Biodynamic Spray and Compost Preparations,* Floris

Klocek, Dennis, *Sacred Agriculture: The Alchemy of Biodynamics,* Lindisfarne

Koepf, H.H., *The Biodynamic Farm: Agriculture in the Service of Humanity,* SteinerBooks, USA

—, *Koepf's Practical Biodynamics: Soil, Compost, Sprays and Food Quality,* Floris

König, Karl, *Social Farming: Healing Humanity and the Earth,* Floris

Kranich, Ernst Michael, *Planetary Influences upon Plants,* Biodynamic Association, USA

Lepetit, Antoine, *What's so Special About Biodynamic Wine?* Floris

Masson, Pierre, *A Biodynamic Manual,* Floris

Morrow, Joel, *Vegetable Gardening for Organic and Biodynamic Growers,* Lindisfarne

Osthaus, K.-E., *The Biodynamic Farm,* Floris

Pfeiffer, Ehrenfried, *The Earth's Face,* Lanthorn

—, *Pfeiffer's Introduction to Biodynamics,* Floris

—, *Weeds and What They Tell Us,* Floris

—, & Michael Maltas, *The Biodynamic Orchard Book,* Floris

Philbrick, John and Helen, *Gardening for Health and Nutrition,* Anthroposophic, USA

Philbrick, Helen & Gregg, Richard B., *Companion Plants and How to Use Them,* Floris

Sattler, Friedrich & Eckard von Wistinghausen, *Growing Biodynamic Crops,* Floris

Selg, Peter, *The Agricultural Course: Rudolf Steiner and the Beginnings of Biodynamics,* Temple Lodge

Steiner, Rudolf, *Agriculture (A Course of Eight Lectures),* Biodynamic Association, USA (also published in another translation by Rudolf Steineer Press, UK)

—, *Agriculture: An Introductory Reader,* Steiner Press, UK

—, *What is Biodynamics? A Way to Heal and Revitalize the Earth,* SteinerBooks, USA

Storl, Wolf, *Culture and Horticulture,* North Atlantic Books, USA

Thun, Maria, *Gardening for Life,* Hawthorn

—, *The Biodynamic Year,* Temple Lodge

Thun, Matthias, *Biodynamic Beekeeping,* Floris

—, *When Wine Tastes Best: A Biodynamic Calendar for Wine Drinkers,* (annual) Floris

Weiler, Michael, *The Secret of Bees: An Insider's Guide to the Life of the Honeybee,* Floris

Wright, Hilary, *Biodynamic Gardening for Health and Taste,* Floris

Biodynamic Associations

Demeter International
www.demeter.net
Australia:
Australian Demeter Bio-Dynamic
demeterbiodynamic.com.au/
Biodynamic Agriculture Australia
www.biodynamics.net.au
Canada (Ontario): Society for Bio-Dynamic Farming & Gardening in Ontario
biodynamics.on.ca (see also USA)
India: Bio-Dynamic Association of India (BDAI)
www.biodynamics.in

Ireland: Biodynamic Agriculture Association of Ireland
www.biodynamicagriculture.ie
New Zealand:
NZ Biodynamic Association
www.biodynamic.org.nz
South Africa: Biodynamic Agricultural Association of Southern Africa
www.bdaasa.org.za
UK: Biodynamic Association
www.biodynamic.org.uk
USA: Biodynamic Association of North America
www.biodynamics.com